WITHDRAWN

How to Succeed in Academics

How to Succeed in Academics

Second Edition

Linda L. McCabe
and
Edward R. B. McCabe

UNIVERSITY OF CALIFORNIA PRESS

Berkeley   Los Angeles   London

University of California Press, one of the most distinguished university presses in the United States, enriches lives around the world by advancing scholarship in the humanities, social sciences, and natural sciences. Its activities are supported by the UC Press Foundation and by philanthropic contributions from individuals and institutions. For more information, visit www.ucpress.edu.

University of California Press
Berkeley and Los Angeles, California

University of California Press, Ltd.
London, England

Library of Congress Cataloging-in-Publication Data

McCabe, Linda L.
How to succeed in academics / Linda L. McCabe and Edward R.B. McCabe. -- 2nd ed.
    p. cm.
    Includes index.
    ISBN 978-0-520-26268-3 (pbk. : alk. paper)
    1. College teachers. 2. College teaching—Vocational guidance. 3. Career development.
4. Universities and colleges—Faculty. I. McCabe, Edward R. B. II. Title.
LB1778.M27 2010
378.1'2023—dc22   2009050654

Manufactured in the United States of America

16 15 14 13 12 11 10
10 9 8 7 6 5 4 3 2 1

The paper used in this publication meets the minimum requirements of ANSI/NISO Z39.48-1992 (R 1997) (*Permanence of Paper*). ♾

This book is dedicated to our mentors (past, present, and future), our parents, our trainees and our children. We thank you for all that you taught us in the past, and look forward to learning even more from you in the future.

CONTENTS

PREFACE

Our goal in writing this book is to spare you the difficulties that we experienced in our combined academic experience of more than seventy years. We also wanted to provide a basic text that can be used in a course meeting the requirements of the National Institutes of Health Office of Research Integrity for ethics training of graduate students and postdoctoral fellows.

Looking back on the first edition, we realized that there had been many changes due to advances in computer technology and the Internet, that we had not included some important topics, and that we had learned a great deal from our own misadventures and the tribulations of others. We also realized that while many of us have careers that are not the typical tenured professor instructing students and pursuing grant funding, our jobs still have many aspects of an academic career.

In both editions we have attempted to be gender neutral. We view sexual harassment as another form of power abuse. We have seen examples of all possible combinations of genders for abusers and abused in these situations. All are reprehensible. We believe that individuals of any gender can experience power abuse, and we want to help everyone deal with these issues.

Like many baby boomers, we were part of the first generation in our families to go to college. We had no models in our families for the situations we encountered. This is why our mentors were our lifeline.

Since we wrote the first edition, mentoring has become a cliché. However, calling oneself a mentor and being a mentor are two very different things. For some, meeting the minimal requirements of the program (e.g., two meetings a year) constitutes mentoring. Many who want the title and benefits of mentorship have absolutely no intention of making the personal sacrifices necessary to put their mentee's interests above their own. Still other alleged mentors view the role as an opportunity to exploit their mentees. We hope that mentees who are encumbered with such "mentors" will extricate themselves from their predicament with the assistance of this book and some real mentors.

When trainees or junior faculty members are confronted with a problem, they may feel that there is something wrong with them and that they have created a situation too embarrassing to discuss even with a trusted mentor. The mentor's role is not to judge, but to help the mentee deal with their challenge and reach the most positive outcome possible. You should have a mentor for each stage and aspect of your career. In the absence of a mentor or mentors who can address each topic and issue, we hope this book will help to provide guidance for you.

How to Succeed in Academics was originally intended for postdoctoral fellows and junior faculty members, but undergraduates, graduate students, and more senior faculty members have participated in our workshops or read the first edition and have found these to be helpful. Whether you are reading this as part of a group or as an individual, we hope the example situations in the vignettes provide insight into problem-solving strategies. It is clear from our workshops that professionals find it helpful to know that others have similar challenges and are working toward solutions.

Although excellent mentors may not receive the recognition that funded grants, publications, and patents bring, their contributions in the form of their mentees will be longer lasting, as their mentees train the next generation.

CHAPTER 1

Choosing a Mentor

Having a mentor is the key to success in academic life. Choosing the right mentors and knowing when to go to them for advice is crucial. We all make mistakes, but we need to let our mentors help us learn from these mistakes and move on with our lives and our careers. While there are no do-overs, there are opportunities to excel, to make wise choices, and to make the most of a bad situation. Just as none of us can do everything well, you may need more than one mentor to cover all aspects of your academic career. Your mentor should not be your friend. Your mentor should challenge you to reach higher, work harder, and go farther than you dreamed possible, at the same time making you aware of the realities of academic life. In seventy years of combined experience in academia, we've made our share of mistakes. That said, an academic life is very rewarding, and the best part is being a mentor yourself. Some of our best moments occur when a current or former trainee or junior colleague asks for advice, shows insight, or shares a success. The best way to pay your mentors back is to be the best mentor you can be!

Choosing a Research Mentor

Lee chose Dr. Johnson as her research mentor because Dr. Johnson was well published and was rumored to be in contention for a Nobel prize. Lee wanted the stimulation of such an exciting research environment. As a postdoctoral fellow, she was assigned to work on a project with two PhD postdoctoral fellows and two junior faculty members. Over three years, Lee was given only menial tasks and for her efforts received only middle authorship on the major paper produced by her group. When looking for a research mentor, Lee should have met with trainees in each of the groups in which she had an interest. She should have determined if the trainees had their own research project for which they were solely responsible for the work and would be given first authorship on any publication. While a major paper is prestigious, middle authorship is not, and directing your own project and having first authorship are more important.

QUALITIES OF A GOOD MENTOR

The most important quality of a good mentor is the ability to take the "me" out of "mentor." In other words, a mentor should put the interests of the mentee above their own self-interest. Each of us owes a great deal to those who have helped us along the way. Unfortunately, not everyone recognizes their debt to their mentors, or their obligations to their mentee.

Your Mentor Should Have
Your Best Interests in Mind

Chris has just started as a new assistant professor. His mentor, the department chair, is constantly providing "opportunities" for Chris. These include a heavier teaching load than anyone in the department, membership on several time-consuming committees, and administrative responsibilities within the department. Chris worries that these activities take time that should be spent on research or creative projects. He worries that when the time comes for a promotion and tenure decision, it will not be favorable. Assistant professors in other departments seem to teach less, participate on only one committee, and have no departmental administrative responsibilities. When Chris confronts his department chair, he is told not to worry. Not reassured, he speaks to a senior professor in another department, who tells Chris that his predecessors were all denied promotion and tenure due to lack of research and creative activity. Chris and the senior professor meet with Chris's chair. The three of them agree on a reduction in Chris's teaching load and committee and administrative commitments so he can devote more time to his research/creative activity.

A mentor should have personal integrity. All personal confidences from the mentee should remain confidential, unless the mentee gives the mentor permission to share information with a particular individual. The mentee should be comfortable sharing successes and problems with their mentor, knowing that the mentor will maintain confidentiality.

A positive outlook on the part of the mentor is helpful. Typically, mentees seek out their mentors for advice when things are not going well. Mentors need to respond with support and encouragement and not complain about their own problems.

The mentor's ability to provide support and training in your chosen field is important. Your mentor should be successful in their academic career and should have completed appropriate training. They should provide training in principles, judgment, and perspective, in addition to research skills. They should introduce you to colleagues in your field.

Appropriate Mentoring Behavior at Professional Meetings

Kelly was a new member in the research group and was pleased to be presenting at a national professional meeting. In addition to practicing the presentations of each member of the research group, the mentor discussed expectations at the meeting. All members of the group would attend the entire sessions in which group members presented. All members of the group would attend the social events of the meeting, including the opening mixer, the department's reception, and the research group's annual dinner for current and former members. The mentor explained that each group member should support the other members during their presentations, and provide opportunities for the mentor to introduce group members to other colleagues and develop their own professional network.

Mentors should provide opportunities for teaching, reviewing manuscripts and grant proposals, serving on committees, and developing leadership and mentoring skills.

Choose a Mentor Who Is Interested in Your Future

Kim was thrilled to be selected to join the highest-profile group in his training program. His mentor was expected to win the most coveted prize in their discipline. What Kim didn't know was that the prize would come from the desperately hard work of the trainees, many of whom would not receive any credit for their efforts. Kim was assigned a repetitive task that was a small part of a very large research program. There was no independent project for Kim to develop, nor would he gain experience in writing grants and manuscripts, presenting at professional meetings, or sharpening any other skills. Kim's mentor did win the prize while Kim was a trainee. When it came time for Kim to apply for his next position, his advisor was too busy being celebrated to send letters of reference or to speak to colleagues about Kim. In updating his curriculum vitae, Kim found little to add from this training period. Kim's mentor might or might not have won the prize without Kim's hard work. It is certain that Kim had been used by an advisor who was not a mentor and who had no interest in Kim's career.

Creative problem solving is an important characteristic of a mentor. This includes helping the mentee to focus, set short-term and long-term goals, and measure progress. The mentor can help the mentee understand that assessing progress may require looking back one year or as many as five years. A long-term goal, such as promotion and tenure, may seem quite vague. We all have a tendency to procrastinate and expect that we can work hard at the last minute and succeed. This does not work with promotion and tenure, where you are being judged on your whole career and your regional and national reputation. A mentor can help the mentee understand what is required for promotion and tenure and how to break that long-term goal into several short-term goals that are easier to visualize and to achieve. These shorter-term goals should be reviewed at least annually to ensure progress toward the long-term goal. By having more frequent meetings, the mentor can help the mentee develop better strategies for success if the mentee is falling behind in attaining short-term goals.

Career Development Takes Time

Kim is a physician who is beginning a postdoctoral fellowship to receive training in a clinical subspecialty and in research. Her long-term goal is to be a "triple threat"—a physician/researcher/educator in a medical school. Having acquired clinical skills during four years of medical school and three years of a specialty residency, Kim is very comfortable taking care of patients. Entering a research setting for the first time has totally perplexed her. All the other members of the research group are comfortable with their projects, methods, and abilities. Kim's initial research attempts lead to one failure after another. She meets with her mentor to reconsider her decision to pursue research. Her mentor asks Kim to consider how long she spent developing her clinical skills..The mentor notes that many of the postdoctoral fellows have had seven or more years doing research—equivalent in time to her clinical training—and everyone in the group was a research neophyte at one time. Kim just needs to put as much effort into research as she has put into mastering clinical medicine.

To establish a national reputation, one needs invitations to speak at other academic institutions around the country and at national meetings, publications in nationally recognized journals, service as a reviewer for nationally recognized journals, book authorship or editorship, committee membership for national professional organizations, and/or job offers from other academic institutions (see table 1).

ADDITIONAL QUALITIES OF A GOOD MENTOR

Mentors should help mentees identify their strengths and weaknesses and learn how to deal with their weaknesses. They should also help mentees deal with institutional realities. Mentors should complement their mentees. If a mentee is struggling to balance a career with a new baby, a mentor with multiple children and a successful career might be appropriate. A mentee who is dedicated to teaching to the point of excluding all other professional endeavors would do well to have a mentor who wins teaching awards and is the best-funded researcher in the department. A mentee who would like to dedicate herself to institutional and professional society committee work needs a mentor who believes there was never a committee worth the time of a junior faculty member. The mentee who lacks focus and is unable to complete any task should have a mentor who excels in time management and will not accept the mentee's excuses. Such mentors might not be selected by the mentees who need them and may have to be appointed by the department chair.

One aspect of an academic career often difficult for young people to achieve is focus. As part of identifying goals, a mentor can help a mentee to select an area of scholarship. The mentor then guides the mentee in the acquisition of expertise on this topic. Mentees establishing their professional reputation in this area need to understand

Table 1. *Timeline of Activities for Establishing a National Reputation*

| | \multicolumn{6}{c}{Year} | | | | | |
	1	2	3	4	5	6
Scholarly activity and research	✓	✓	✓	✓	✓	✓
Write articles for peer-reviewed journals	✓	✓	✓	✓	✓	✓
Apply for membership in regional professional organizations	✓	✓	✓			
Submit abstracts for presentation at regional meetings	✓	✓	✓			
Volunteer for committee work in regional professional organizations		✓	✓			
Apply for membership in national professional organizations				✓	✓	✓
Submit abstracts for presentation at national meetings	✓	✓	✓	✓	✓	✓
Volunteer for committee work in national professional organizations				✓	✓	✓
Apply for grants	✓	✓	✓	✓	✓	✓
Volunteer to serve on a grant review panel						✓
Review articles for journals	✓	✓	✓	✓	✓	✓
Write review articles for peer-reviewed journals				✓	✓	✓
Write chapters in major textbooks				✓	✓	✓
Volunteer to serve on an editorial board						✓
Edit a book						✓
Write a book						✓
Apply for positions at other institutions						✓

that they need to be very careful. While it takes years to establish a reputation, one misstep can ruin a career. Academic professionals need to trust their colleagues, and when this trust is threatened, the effect can be devastating.

Consulting and clinical work can bring more rapid, wider, and more remunerative recognition than research. Rewards for teaching and curriculum development may be less concrete. Support for the academic infrastructure through committee work and administration can be time-consuming and may not be recognized. However, each of these activities is essential to the development of your academic career and to the support of your institution and your discipline.

A good mentor is measured by the success of their mentees. Good mentors encourage excellence and scientific integrity on the part of their mentee. If you are considering mentors, look at the first authors of prior publications of potential mentors. Have these first authors gone on to develop independent careers?

Celebrate a Mentee's Success

Dr. Jones was the research mentor for both Kelly and Stacey. Kelly was promoted to faculty at the mentees' original institution, while Stacey left for a faculty position at another institution. Dr. Jones left to chair a department a third institution. She sought funding for an endowed chair for her position meaning that the donation would be invested and would support Dr. Jones and her successor chairs. Kelly called Dr. Jones to say that he had received an endowed chair. Dr. Jones was proud of Kelly, and the fact that her mentee received an endowed chair before she did was unimportant. What was important was that Kelly's abilities had been recognized by a donor. Stacey soon e-mailed Dr. Jones to invite her to attend the celebration of Stacey's own endowed chair. Dr. Jones was honored to be included and enjoyed the opportunity to celebrate Stacey. When Dr. Jones obtained funding for her endowed chair, Kelly and Stacey joked that they had taught her a thing or two about getting an endowed chair. Dr. Jones replied that she still had a lot to learn and was glad to have Kelly and Stacey to teach her.

Mentors not only are concerned with the current stage of their mentee's career but are also proactive, encouraging their mentee to develop an independent career path. They should commit to assisting the mentee in making the next move in their career. This involves providing opportunities for the mentee to write (e.g., abstracts for professional meetings, manuscripts, grant proposals) and to present their work at professional meetings. Mentors should nominate the mentee for awards, help the mentee plan for their next career move, and serve as a reference for the mentee.

Breaking Up Is Hard to Do

Dana excelled as a graduate student and was flattered when his mentor asked him to stay on as a postdoctoral fellow to continue his project. Dana applied to a number of funding sources within and outside of the institution for support of his fellowship, and was shocked when each request was denied. He did not understand that funding sources were looking for independence from the graduate school mentor, not just more of the same. The funding sources were actually looking out for Dana's best interests, encouraging him to develop his own independent line of research.

Junior faculty members should have a mentor dedicated to helping them achieve promotion to associate professor and attain tenure. This mentor should be outside the section, division, or department so they will have no conflict of interest and be interested solely in the faculty member's success in the tenure track. Each institution has its own rules for promotion and tenure. Some general criteria include a national reputation, creative products (e.g., publications and inventions), quality of teaching, and committee service.

Alleged Independence

Robin, having worked hard as a fellow in Dr. Stevens's laboratory, had just moved to another school to be an assistant professor. He planned to continue one of the projects he had developed as a fellow. Dr. Stevens called Robin weekly to check on his research progress and to discuss his results. Robin appreciated Dr. Stevens's interest in his research. When Dr. Stevens asked him to send his raw data to her electronically, he did not hesitate. A few months later, as Robin was preparing an abstract on his research for a meeting, a colleague came by with the latest issue of a major journal to show Robin an article with his data and Dr. Stevens as senior author. Clearly, Dr. Stevens took advantage of her relationship with Robin and did not encourage Robin's independence. Even if Dr. Stevens included Robin as a coauthor, her behavior would be inappropriate. At this point, Robin should seek advice from his other mentors at his former institution, as well as his current mentors. With these advisors, Robin should confront Dr. Stevens and demand that she submit an erratum to the journal, making Robin the first or senior author, depending on how much data other than Robin's was included in the paper. If Dr. Stevens is unwilling to do this, Robin and his mentors should contact the journal editor and ask for assistance. The editor will contact Dr. Stevens and demand that the authorship be revised and that letters from each of the authors and Robin be provided to the editor, indicating agreement with the revised authorship. The editor will then print an erratum with the new authorship.

True Independence

When Kerry enters Dr. Johnson's group as a junior faculty member, they agree to collaborate on one topic and allow Kerry to develop a second topic independently. After three years, Kerry has several publications and some independent funding and is ready to start her own group. She moves into independent space and recruits her own trainees.

She continues to collaborate with Dr. Johnson on their joint project, and both are authors on the papers that result. Kerry continues to work independently on her other project. While Dr. Johnson is available to discuss ideas, read rough drafts of manuscripts and grants, and help with priorities and goal setting, she is not a co-author on these papers or a collaborator on the grants. If Dr. Johnson is invited to speak on this topic, she refuses and insists that Kerry is the expert.

SELECTING A MENTOR

Selecting a mentor for one particular characteristic is never a good idea. You wouldn't buy a car just because it had the engine you wanted; rather, you would consider all the attributes of the vehicle. The same is true of mentors. We feel that everyone, regardless of gender or ethnocultural group membership, has the same issues. We have seen trainees select mentors based on a single characteristic, such as ethnocultural group membership or gender, and discover that they made a big mistake. Examine the entire package, not just one characteristic, when you search for a mentor.

Selecting a Mentoring Committee

Susie was concerned about gender bias. While her research mentor was male, she wanted the rest of her graduate school mentoring committee to be female. She selected women for her committee without considering their mentoring track record. Imagine her dismay at her committee meetings when her "sisters" were arbitrary, inconsistent, and demanding. At Susie's dissertation oral, these women got together ahead of time and planned to ask questions that would embarrass Susie. Fortunately, Susie's male research mentor was able to move the committee to consensus, and Susie received her degree. Susie vowed to be gender neutral in her future selection of mentors.

Another bias that trainees sometimes use in selecting mentors is preference for a mentor at one or another stage of their career. Some want a young mentor, fresh out of postdoctoral training, who will be "hands on." Unfortunately, this person may have no track record as a mentor and may or may not develop these skills. Other trainees are seduced by the resources and clout of senior mentors. While these individuals may have been successful in the past, it is important to determine if they are still committed to mentoring.

HOW DO YOU KNOW IF SOMEONE IS A BAD MENTOR?

Bad mentors are serial abusers. They "eat their young." They abuse their colleagues and collaborators. They may insist that trainees repeat experiments until they get the "right answer," the one that agrees with the mentor's hypothesis. They may prolong the mentee's training program by switching the trainee from one project to another. They may encourage the trainee to take a leave of absence, a terminal masters degree, or leave the program. Each of these choices makes the mentee's lack of progress look like the trainee's fault. If the trainee succeeds in defending their dissertation, the abusive mentor may block publication of the mentee's research. If you find yourself with a bad mentor, ask your committee, the head of your program, the department chair, or the institutional ombudsperson for help. At least one of these individuals is probably aware of the past and current abusive behavior and will be willing to help you develop a creative solution to your problem. The ombudsperson receives and investigates complaints and seeks equitable resolution of disputes. Such resolution may require changing mentors. While this change may cost time, you need a mentor who will guide you through your training and launch you on the next step of your career.

WHICH ASPECTS OF YOUR
CAREER REQUIRE MENTORING?

All aspects of your academic career require mentoring. Each aspect requires a particular skill set, and help from your mentor in adopting successful strategies is crucial. One aspect of your career that used to be left out of the mentoring equation is the mentee's personal life. However, whenever we give our How to Succeed in Academics workshops, we are asked questions more and more regarding how to balance personal and professional lives. This involves learning time management skills.

ONE MENTOR IS GOOD;
MORE THAN ONE IS BETTER

It is hard to imagine that one mentor would suffice for all the different aspects of your career. You will need different mentors at different stages and for different aspects of your career. As you look for your first position, you will need mentoring for interviewing and negotiating a contract. As you take on new responsibilities such as teaching, you may need new mentors to help you learn and evaluate teaching skills. Research is so complex, often crossing traditional departmental boundaries, that you will often need more than one research mentor.

Establishing Ground Rules

Dr. Jones and Dr. Smith met at a seminar. Each was each intrigued by the questions the other asked of the speaker, and they decided to continue their conversation. They developed a collaborative project, and Dr. Jones had a new graduate student, Kerry, who was interested in pursuing this project for his dissertation. Before they started, Kerry, Dr. Jones, and Dr. Smith met, agreed that this would be Kerry's project, and worked out how it would be funded. Kerry would be responsible for carrying out the research under Dr. Jones's and Dr. Smith's supervision. Dr. Jones and Dr. Smith would conduct weekly group meetings to discuss progress with Kerry. Kerry would be responsible for writing the manuscripts and would be the first author. Dr. Jones and Dr. Smith would alternate senior authorship on any publications resulting from this collaboration.

We encourage trainees to have mentors outside their division, section, and department. This decreases conflicts of interest, increases the likelihood of independent research careers, and increases diversity within the group.

Lack of Diversity of Research Interests within the Department

Dr. Smith is the chair of a small department. Each faculty member is involved in Dr. Smith's research program. No other research is done in the department. Every postdoctoral fellow and graduate student is involved in Dr. Smith's research program. Each publication identifies Dr. Smith as the senior (last) author and the faculty member responsible for the work as the first author. Any postdoctoral fellows or graduate students involved are middle authors. Dr. Smith is the principal investigator on every grant, so he controls research funding. How can the faculty members, fellows, and students develop independent career paths? What research credentials do they have if they look for a position outside the department? What happens if Dr. Smith's area of research is no longer fundable or publishable?

Such a lack of diversity of research interests is clearly a conflict of interest (see chapter 14) and has numerous adverse consequences, particularly for the trainees. The requirement that mentoring committees monitor training programs can help prevent such abuses, unless the other members are all part of the same research group.

These committees should not provide mere window dressing for the purpose of an application for support. They need to meet regularly with each trainee. Trainees should also take advantage of the larger training context of a program, center, department, or school, since this provides additional mentors. Continued funding of these training programs is based on the trainees' career success, so those administering the program are invested in the success of their trainees.

Establishing the Time Commitment

Stacy was a new assistant professor in gastroenterology. Stacy had negotiated 75 percent time for research in accord with the requirements of his K08 grant. Dr. Smith, Stacy's division chief, called Stacy into her office to discuss Stacy's responsibilities to the division. Dr. Smith showed Stacy the outpatient clinic schedule and the call schedule. When Stacy added it up, clinical responsibilities would be 35 percent of Stacy's effort. Then Dr. Smith suggested that Stacy should fill a department slot on the medical school admissions committee. When Stacy asked how much time that would require, Dr. Smith replied ten to twenty hours per week, but only for six months of the year. She then recommended that Stacy assume responsibility for the weekly division noon conferences. Clearly Stacy needs a mentor outside the division who can remind Dr. Smith that accepting an NIH grant means not only accepting the financial support for Stacy but also guaranteeing Stacy the 75 percent of her time required by the grant for research. Stacy cannot be distracted by too much clinical work and committee responsibilities.

Having multiple mentors is invaluable if the mentee and main mentor have problems. The mentee should have access to other individuals with whom to discuss these issues. If a mentor is being unreasonable, the mentee needs to discuss this with the mentor. It is helpful to keep relevant e-mails and other written material. The mentee should also seek advice from other mentors, especially those without a conflict of interest. If the mentee has a mentoring committee, director of training, or department chair, the mentee should consult them. In addition, most institutions have an ombudsperson to assist with difficult negotiations. If all of these possibilities fail, the mentee may need to seek legal assistance. Regrettably, some institutions are not responsive to trainee needs and have a culture that tolerates what should be intolerable behavior on the part of supervising faculty (we choose not to call them mentors). Lawyers can get their attention.

Having Multiple Mentors

Before Kim moved to accept a faculty appointment in endocrinology, she identified a research mentor in the Department of Biochemistry. Upon arriving, she was surprised when her department chair, Dr. Johnson, assigned her a mentor for promotion and tenure outside the Division of Endocrinology. Dr. Johnson explained that the department wanted to ensure Kim's success with promotion and tenure by providing a mentor who has a focus on her career and no conflict of interest (see chapter 14).

YOU NEVER OUTGROW YOUR NEED FOR MENTORS

Each aspect of your career requires mentoring. If you decide on an administrative career path, you should explore the possibilities with those in such a position as soon as possible. You need to understand the qualifications, duties, and goals of the position. Individuals who hold such administrative positions will be asked to recommend candidates for such positions.

While it may be difficult to recognize this if you are early in your career, you may be considering retirement in the future. Talking to those who have successfully transitioned into retirement may be very helpful. Everyone has a unique solution to their life change, but it is always helpful to learn from the experience of others.

MENTORING IS A TWO-WAY STREET

Getting Started

Although Chris had never done research before, she was excited by the study that Dr. Roberts had carried out. Chris knew she had a lot to learn. Dr. Roberts assigned Chris to a senior PhD postdoctoral fellow for training. Chris and the PhD worked well together, but Chris found that, while the postdoctoral fellow was busy with something else, she had time to learn techniques from different members of the group. Chris offered to help others in the group when she could. She is learning new techniques, is showing she is a team player and is establishing her place in the group. She is proving to be a valuable asset while increasing her own knowledge base. Experience has taught us that the best predictor of success is fearlessness, a willingness to take on new challenges.

Even the best mentor cannot compensate for lack of response from the mentee. Mentee and mentor need to engage in open communication. They need to meet frequently, not simply fulfill the minimal requirement to prove that they are in this relationship.

Dealing with the Reluctant Mentee

Dr. Smith has a successful track record with mentees, but is frustrated by a lack of interaction with Lee. When Lee joined her group, Dr. Smith carefully explained that Lee would be expected to attend weekly two-hour group meetings, where Lee would update the group on his progress, occasionally be responsible for journal club, and assist group members with their research. In addition, Dr. Smith specified that Lee needed to meet with her for at least a half hour every week for more specific research direction and to discuss any issues Lee was having with the training program. In fact, Lee came to group meeting only about half the time and never made an appointment to meet with Dr. Smith in spite of her encouragement to do so. Dr. Smith eventually confronted Lee at his mentor committee meeting. *The other members of the committee agreed with Dr. Smith and urged Lee to attend all weekly group meetings and to meet with Dr. Smith weekly. In spite of these suggestions, Lee steadfastly refused to interact with Dr. Smith. When Lee did not pass his qualifying exams and received a terminal master's degree, Dr. Smith was very disappointed.*

If in spite of the mentor's best efforts, if the mentee is not engaged, the mentor should confront the mentee and remind the mentee of their responsibilities. If the mentee is still not responsive, the mentor should convene a meeting with the mentee and the mentee's mentor committee for a tough love session. It is better to intervene early in the mentoring relationship than to wait until later. The lack of responsiveness may simply be a clash of personal styles. It may be that the mentee would do better with a different mentor. The mentor may have to accept that the mentee's other needs may conflict with the mentee's academic success.

REWARDS OF BEING A MENTOR

Mentoring is one aspect of your career that will have a tremendous impact ten, twenty, and even thirty years from now. Is there any higher compliment than having someone say that they would like to learn from you or to be like you? We treat our most junior trainees (high school and college students) as collaborators. Your students and trainees can provide tremendous insight into your strengths and weaknesses and can give you outstanding advice.

Learning from Your Mentees

Dr. Jones was a meeting nerd. She came to a professional meeting with a full schedule for each day. Meals were determined by what was available at the venue. After the sessions were over for the day, she reviewed the day's experiences and her plans for the next day. Lee, one of Dr. Jones's fellows, took her aside and said, "I think you have meetings all wrong. Meetings are all about 'face time,' networking, discussing new ideas, and establishing new collaborations." Struck by Lee's insight. Dr. Jones was grateful to him. From that point on, she balanced sessions and interpersonal interactions. She organized a group dinner for current and former trainees at each professional meeting. She contacted colleagues before the meeting to establish time to discuss their joint projects. She attended each meeting's social functions and found that these were good places to renew acquaintances and to introduce her trainees.

DEVELOPING YOUR MENTORING SKILLS

The phrase "see one, do one, teach one" is an archaic approach to medical education but is still very popular in our research group. We will always remember the undergraduate who expressed his gratitude for how much he was learning from his research experience. We replied that he would begin to be a mentor himself when he taught one of the methods he was using to the new high school student.

A mentoring relationship is truly reciprocal. The mentor often learns as much from the mentee as the mentee learns from the mentor. The mentor should provide opportunities for the mentee to learn mentoring skills. This may take the form of assigning the mentee to supervise and train a new member of the group. This may become more formalized when the mentee shares mentoring responsibility with the mentor for a trainee. The mentor may then gradually decrease their responsibility for the trainee and allow the more senior mentee to take over. We call this mentoring the mentors.

The mentor should demonstrate the value of mentoring and praise the more senior mentee for showing appropriate mentoring behavior.

CHAPTER 2

Preparing for Graduate
or Professional School

CHOOSE A TRAINING
PROGRAM CAREFULLY

How do you decide which training program to pursue? It is worthwhile to spend a lot of time and personal energy in this decision process. If you are fortunate to be certain of a particular training program, you should still take the time to consider why you have chosen your career path. This will be one of the main questions on application forms, in interviews with your professors who will serve as references for you, and in interviews with program faculty members and trainees. Simply stating "I've wanted to do this as long as I can remember" rings hollow. If the training will enable you to meet your goals of developing your talent, helping others, enabling you to better express yourself, expanding and sharing your knowledge base, and/or providing the tools to solve problems, you have a rational basis for pursuing this training.

Medicine and Science Offer Many Career Opportunities

Sandy's parents are both physicians. Growing up, whenever Sandy was asked, "What do you want to be when you grow up?" she always replied, "A doctor like Mommy and Daddy." She was always bandaging dolls and pets and running a "hospital" for them. Sandy took all possible science courses in high school and applied only to colleges that provided a strong premedical education. When Sandy entered medical school, she felt a great deal of pressure to devote all of her attention to medicine. Sandy was conflicted because she had always enjoyed art as well as science. There had been time for both art and science, but medical school demanded a higher level of commitment. Worried that medicine would cause her to abandon art, Sandy determined that she could combine both as the writer and illustrator of medical texts.

QUESTIONING YOUR DECISIONS IS NORMAL

You should recognize that you will intermittently question your career path. Any graduate or professional degree program is long and arduous, and at times your interest may seem to ebb. At such moments, don't make any quick decisions to dramatically change your direction. Also, you should recognize that each field has a wide breadth of opportunity and can accommodate individuals with diverse interests.

SUCCESS REQUIRES PASSION

To be successful in any endeavor you pursue, you must be passionate about what you are doing. It is difficult to fake passion, but it may be possible to do so during the interview process. The problem will be if in the ensuing years you lack the enthusiasm required to be productive and happy. Be honest with yourself, and ask your mentors to be honest with you, as you go through the decision process.

QUESTIONS TO CONSIDER IN
CHOOSING A CAREER PATH

You should ask yourself questions such as the following to determine which career to pursue:

What do you like to do?

Do you enjoy teaching?

Do you like to work with others or alone?

Can you tolerate delayed gratification, or do you want immediate feedback?

Do you want to travel?

Do you enjoy writing?

Do you want to focus your energy in one area or be involved in a number of areas?

Do you want to be at the center of a field or at the interface between fields?

What do you want to be doing professionally ten years from now?

Are you willing to be in training ten years from now?

What type of setting do you want to be working in ten years from now?

Do you see yourself as a mentor?

When you look back at your professors who have completed a similar training program, would you be comfortable in their roles?

Would you consider leaving academics for the private sector?

Are graduates from this training program able to secure positions in settings you would find interesting?

What level of responsibility do you wish to attain?

Do you see yourself advancing into an administrative leadership role?

What would be your eventual goal?

What level of risk are you willing to assume?

How important is financial security?

How important is job security?

Would the training program mean that you would have to assume a great deal of debt?

Would the training program mean a significant delay before you have a "real" job?

What is the ratio of the number of graduates from the training program to the number of openings you find interesting?

Do graduates move from this training program into jobs or into further training?

How important is independence in your personal equation?

WHICH TRAINING PROGRAMS IN YOUR CHOSEN FIELD ARE THE BEST FOR YOU?

Discuss with your mentors which programs have consistently accepted applicants from your institution. This increases the likelihood of your acceptance if you are competitive. Your mentors can also give you a feel for the quality of the programs and faculty members. Your mentors may also know about the mentoring track record of the training program faculty you consider to be possible mentors.

Search the websites of the institutions and programs you are considering. Be sure that the programs offer what you want by considering the faculty's creative activities, courses offered, requirements to graduate, and the products produced by previous and current trainees. Financial factors include cost of living in the location, tuition rates, book and equipment expenses, and financial aid. You need to be fully informed regarding loan repayment requirements. You don't want your loan repayments to determine your career choices when you are finished training.

Finances May Determine the Future

Kim graduated from medical school with a large debt. She had always wanted to pursue an academic research career in an interventional specialty like cardiology or gastroenterology. However, her partner and her family were placing a great deal of pressure on her to complete her pediatric residency and join a large group practice to make money to repay her debt. Fortunately, Kim spoke with a cardiology fellow who explained that the National Institutes of Health (NIH) had loan repayment programs for fellows and junior faculty pursuing research. Kim also discussed her financial concerns with the gastroenterology division chief, who pointed out that while she might make more money in practice than she would as a fellow, academic salaries eventually outpaced salaries in the community. Kim decided that she could manage the financial issues in the short term and that she would pursue her passion for interventional pediatrics by pursuing a two-year fellowship in critical care, performing research, and seeking a faculty position. Since there is an extreme shortage of pediatric subspecialists, she had a choice of a number of excellent academic jobs.

The information available on the Internet can empower you in your search for a career path, training programs, and mentors. However, it does not replace the very personal information and insight you can obtain from your advisor, professors, colleagues, and trainees (past and present) in a training program or at an institution.

You need to be realistic in terms of the number of programs to which you apply. Applications and interviews (if required) are time-consuming. If you are paying application fees and/or paying for your transportation to interviews, this can be very expensive.

THE APPLICATION PROCESS

Determine what, if any, standardized examinations are required by the programs in which you are interested. If standardized tests are not your forte, you might consider studying for these tests and/or paying tuition to one of the commercial training programs. You could also consider a book or a course that teaches general test-taking skills. Your mentors and your older friends on the same career path can provide advice as to how to proceed. You should take these tests as soon as possible, so you have the opportunity to take them again if you feel that your score is not competitive.

Your mentors can advise you regarding the wisdom of taking the test again. Given the practice effect, you will probably be able to do better the second time. However, if you are too anxious, you need to desensitize yourself by taking practice tests on your own or as part of the commercial training programs. Remember, your best performance comes with a medium level of anxiety, just enough to give you an edge, but not too much to interfere with your performance.

Some testing programs reward you for taking the test more than once. We have been told that the entrance exams for medical school report your best score on each part of the test, regardless of which time you achieved this score. If your best score on part A was the first time you took the test and your best score on part B was the second time, these are the two scores reported.

You also need to be sure that you have the experience necessary to apply. Research or other creative experience is helpful in two ways. It demonstrates that you understand what you are applying for, and it provides a letter of recommendation from your mentor that can specifically describe the personal qualities of yours that will ensure success in the training program. If you are applying in the arts, you need a corpus of work for your application. This needs to be developed over a period of time. Relevant service experience is also required for certain professional schools. Like research, this demonstrates that you understand what you are applying for and provides a letter of recommendation. For medical schools such volunteer service should involve direct patient contact. We recommend to the undergraduates who ask us that they volunteer with a Child

Life–Child Development program. They will work in small groups or one-on-one with pediatric patients. Because the Child Life–Child Development staff closely supervises the volunteers, they provide excellent letters of reference.

The reality is that a letter from one of your professors that says, "This student in my class of two hundred students earned an A," does not mean a whole lot to an admissions committee. You need to cultivate letters from individuals who are leaders in your field of interest by working with them. You should select individuals who are committed to mentoring and who will take the time to prepare a thoughtful letter of recommendation. Writing a letter of recommendation is an art that has to be developed. You also need to provide enough notice to your references so they have time to prepare a thoughtful letter on your behalf. You should meet with them and provide them with your résumé and the forms so they have the materials they need.

By interacting with individuals who have completed this type of training program, you will be able to determine:

If the training program is appropriate for your goals.

If you enjoy interacting with graduates of this type of program—you will spend most of your professional life with your colleagues.

If the training program provides the appropriate skills.

If the mentors in the training program assist in securing the next position.

If there are sufficient positions available after training is completed.

Some schools have an institutional application in addition to the departmental application. Both are required. Some institutions insist that you apply to only one of their departments or programs at a time. Be sure you understand the rules for each institution. The primary application to medical schools is a single, standardized application. However, if the admissions committee requests a secondary application, these can vary from school to school.

Keeping track of deadlines is very important. It is a test of your organizational ability. You should develop a system that works for you so you can meet every deadline and check it off as you do so. This also helps make an overwhelming amount of work manageable. This is especially difficult if you are carrying a full undergraduate course load at the same time. Some students wait to apply until after they have completed their undergraduate degree. Remember life is not a fifty-yard dash, but rather a marathon. If you can use the time between undergraduate and graduate and/or professional school to obtain more relevant experience and focus on your applications, this is time well spent.

You need to check with the office receiving your application to make sure that it is complete and together. You may have to provide material more than once to ensure that your file is complete.

THE INTERVIEW

Some programs require an in-person interview. This can be very helpful if you have researched the program and the institution beforehand, and if you go with specific questions regarding the organization of training and the nature of mentoring. You are judged in your interview by the questions you ask as well as by your answers to questions asked of you. Any discussion should involve approximately equal input by the two parties. Training programs will evaluate your interest and motivation by your interactions during the interview. You must remember that the whole time you are there is part of the interview, including the tour and any meals or social events with the students. While the faculty can answer many of your questions, it is ideal if you can also speak with current trainees to obtain their perspectives. They recently went through the application, interview, and selection process. They can explain why they chose the program they did and discuss whether or not the program is meeting their expectations. They are also very aware of the current prospects for previous trainees or those who have almost completed the program. Be very cautious about any program that does not give you free and open access to its trainees without faculty or staff present.

Be Sure to Meet with Trainees during Your Interview

Kim applied to a training program and was invited for an interview. Her advisor helped her prepare for the interview with a practice session where she was encouraged to ask questions about the training program that were important to her. She met with the program director and several senior faculty members. Kim's interviews went very well. She answered questions for about half of each interview and was able to ask her questions for the rest of the time. At the time, she thought it was strange that there was no interaction with trainees, but the program director and the faculty sold Kim on the program. After her acceptance and relocation to a new city for the program, the grim reality became apparent. The faculty was totally out of touch with reality. They wanted to create clones of themselves, not prepare their trainees to be competitive for positions within or outside of academics. Trainees in this program often extended their training in a vain attempt to become more competitive. Kim was devastated and immediately began to consider transfer to another institution.

QUESTIONS YOU SHOULD ASK DURING YOUR INTERVIEW

During your interview you should ask questions that show you have researched the program and developed specific questions based on your interests. You should not ask the same questions at each program, because every program is different. Your questions should focus on your professional development, not just money and time off. If several of the programs you are interested in are different, you might ask your interviewer to compare and contrast the different programs. Any questions you ask should be nonthreatening and nonjudgmental. You need to be sure that the training program has the flexibility you desire. For example, are there different tracks depending on your interests? You need to be sure that you understand the requirements of the program and how they will shape your career.

Be sure to talk not only with professors but also with current trainees and ask them if the recruitment for the program matches the reality of the training. Ask those close to leaving the program if the program prepared them for the kind of position they want and whether they have been able to secure such a position. You can ask any trainee if there is anything they would do differently if they were to do their training again. You should get a sense of whether or not the trainees are happy with their program.

CHOOSE A PROGRAM FOR ITS STRENGTHS, NOT ITS STATUS

For undergraduates, your choice is between graduate school, professional school, or both (e.g., JD/MPH, MD/MPH, or MD/PhD programs). You should ensure that your personal goals are consonant with the goals of the training program. Make sure that your only goal is not to prove something to someone; that is, choose a program for its strengths, not for its status. The smart people select the program that is most appropriate to their personal goals.

Attempting to Prove Something

Stacey was high school valedictorian, graduating with a 4.56 average, and was approaching the senior year of college with a 3.96 grade point. He had always wanted to be a physician. His parents were both physicians, and both enjoyed taking care of patients in a private-practice setting. Stacey was aware of the rewards and the rigors of medicine and decided that caring for patients was the most gratifying position one could attain. He was a very competitive person and, in applying for medical school, decided to apply to combined MD/PhD programs because he felt the training would be far more rigorous than medical school alone. Typical MD/PhD programs include the first two years of medical school, three or more years of graduate school, and the last two years of medical school. Stacey had never attempted research outside high school and college laboratory coursework. With strong grades, excellent letters of reference from professors, and good scores on the Medical College Admission Tests (MCATs), Stacey was invited to interview by several MD/PhD programs. During the interviews, he could not answer questions such as, What types of scientific questions excite you? How will you combine subspecialty training with research training? What research experience have you had? What qualities are you looking for in a scientific mentor? How will you feel in four years when the students you entered medical school with are graduating with MDs and you have finished two years of medical school and are only completing your second year of graduate school? Fortunately, several of the interviewers suggested that Stacey amend his application to reflect his interest in clinical medicine and to apply to medical school without the MD/PhD program.

IF YOU AND YOUR PARTNER ARE APPLYING AT THE SAME TIME

If you are one member of a couple and the two of you are applying at the same time to graduate and/or professional schools, this will be a very stressful time in your relationship. Think about all the aspects of your lives that may be changing: trainee roles, location(s), financial needs, increased independence, and so forth. To increase the likelihood of your joint success, both of you should research the possible programs. You should focus on applying to programs in locations that include a number of opportunities for both partners.

You and your partner should discuss how you will deal with the possibility of one of you getting your dream opportunity in location

Table 2. *Schedule for Applying to Graduate or Professional Training Programs*

	Time before you enter the training program							
	4 yrs.	3 yrs.	2 yrs.	1 yr.	9 mos.	6 mos.	3 mos.	1 mo.
Begin to consider career path	✓	✓						
Meet with advisor	✓	✓	✓	✓	✓	✓	✓	✓
Plan coursework to prepare for training	✓	✓	✓	✓				
Gain relevant volunteer experience	✓	✓	✓	✓				
Gain research or creative activity experience	✓	✓	✓	✓				
Have discussions with those in the field	✓	✓	✓	✓				
Search Internet for training programs			✓	✓				
Narrow program selection			✓	✓	✓			
Request application				✓	✓	✓		
Take tests for application				✓	✓	✓		
Visit for interviews						✓	✓	
Select program							✓	
Move to program							✓	✓

A and the other having their ideal situation in location B. Are you willing to be apart for the duration of the shortest training program? How will you maintain your relationship?

If you are applying to the same programs, you are competing with each other. It would be very healthy for your relationship to discuss possible outcomes and how you will deal with them. For example, what if one of you is accepted by your joint first choice and the other one is not? Given the overwhelming numbers of superb applicants, this is not a valid measure of future potential. It is simply the best estimate the admissions committee can provide at that moment given the material they have.

CHOOSE A TRAINING PROGRAM THAT WILL FORM THE BASIS OF YOUR CAREER

If you are completing graduate or professional school, choice of further training depends on which type of training will support your goals. You will need to ascertain which program requirements are important for success in the job market. You need to determine which skills and projects will make you uniquely qualified. Sometimes the most marketable individuals are those at the interface of two fields; therefore, you might consider additional training in a different field. You should review the resumes of faculty members in departments that provide the environment you would like. What type of training did they pursue? You need to determine if you will need more than one training situation (e.g., one or two postdoctoral fellowships) to prepare you for a career. You will need training not only in your area but also in the fundamentals of academic life, such as writing, making presentations at meetings, and teaching.

CHAPTER 3

Preparing an Abstract for a Professional Meeting

PRESENTATIONS AT PROFESSIONAL
MEETINGS ARE CRITICAL FOR THE FOUNDATION
OF YOUR NETWORK AND YOUR CAREER

You may need grant support to guarantee funding for your research program and, in some cases, even for your salary. However, the basis for your grant support will be your reputation, which depends upon communicating with others in your field through your presentations at professional meetings and your reports in publications. Only as a trainee or a beginning faculty member do you receive grant funds based on the ideas in your proposal, on your mentor, and on your potential. As your career advances, the basis of your requests for funding will be your preliminary data and your publications, in addition to your ideas. Peer-reviewed abstracts and publications indicate that others have considered your ideas and data and found them to be substantive. Previous success in review of your work for meeting presentations and, more important, publications lends credibility to your proposal and your commitment to see your research program through to review by your peers. You are also establishing your scientific reputation through your presentations and publications. Your proposal is an attempt to market your research ideas, and so is submitting your abstract to a professional meeting. Your presentations at meetings provide the basis for your publications and grant proposals and help to establish your network of colleagues.

KNOW THE GOALS FOR YOUR PRESENTATION AS YOU PREPARE YOUR ABSTRACT

Your goal in preparing your abstract is to present your ideas to your colleagues. To do this, your abstract must be selected for presentation. Once your abstract has been selected, you also want it to attract an audience. Each of these goals is part of developing respect for your work among your colleagues.

TO BE EFFECTIVE, GOOD SCIENCE REQUIRES GOOD WRITING

The most important requirement for a successful abstract is good science. However, good writing, appropriate for the meeting, is also important. Sometimes, good writing may give the impression that there is more substance to the science than there is. You should be cautious not to promise more than you have in hand, and thus more than you will be able to deliver.

Replication Is the Basis of Science and of a Good Abstract

Kim worked very hard to complete a series of experiments in order to meet the deadline for abstract submissions for the national meeting. One of her goals was to have a presentation at this meeting. When she asked her mentor, Dr. Johnson, if she should prepare an abstract for the meeting, Dr.

Johnson suggested Kim wait for the next meeting. Kim was disappointed, but Dr. Johnson explained that the experiments needed to be repeated. A number of Kim's figures were not convincing. There was the possibility that these data could not be replicated. This would cause Kim to have to withdraw her abstract and would embarrass Kim and Dr. Johnson. This would not be good for either of their professional reputations. Kim had to agree that Dr. Johnson was right.

A SUCCESSFUL ABSTRACT IS FOCUSED AND UNDERSTANDABLE

There should be one idea or, occasionally, two very closely related ideas that you can sell as one, in your abstract. More than one idea can be used to develop multiple abstracts. You should never use one idea as the basis for more than one abstract. In addition to being focused, a successful abstract is clearly written without excessive jargon or abbreviations. It should be understandable in a single read by someone who is unfamiliar with your work and who may have more than a few hundred abstracts to review within a few days.

A Good Abstract Must Have One Purpose, No More and No Less

Lee is a postdoctoral fellow and Lynn is a medical student. They worked together on the same successful project in Dr. Smith' s laboratory over the summer. Dr. Smith wanted the work presented at a national meeting and wanted both Lee and Lynn to have the opportunity to present their work. Dr. Smith split the single project into two abstracts. One abstract had Lee as first author and potential presenter, and the other had Lynn in those roles. The abstract reviewers were not impressed with these "least minimal presentation unit" abstracts. Lee's abstract was not accepted for presentation, and Lynn's abstract was accepted as a poster.

Dr. Smith then realized that one abstract should have been submitted with the request that it be presented as a poster. Then both Lee and Lynn could have presented their work. When Dr. Smith realized this error in judgment, Dr. Smith asked them to present the poster together.

WHO REVIEWS YOUR ABSTRACT?

Typical abstract reviewers are busy, senior colleagues. They may have only a few days to review two hundred abstracts. Since they have their own work to do as well, they cannot devote full time to abstract review. While they have been assigned to review abstracts on the general topic because of their work in the area, they probably are not in your very specific field. A typical abstract reviewer will look at the title and the data. If these are not clear, not substantive, not interesting, or appear to be an advertisement for a product or service, the abstract may be placed in the "do not select for presentation" pile. If the title and data suggest promise, the reviewer will consider the whole abstract. Exciting and substantive abstracts will be selected for platform presentations. Less interesting abstracts will be selected for posters. The abstract reviewer does not have time to go back and reconsider the abstracts in the not-for-presentation pile.

If the reviewer is at your institution, they cannot review your abstract. They will note their conflict of interest as their score for your abstract.

WHAT CRITERIA DO ABSTRACT REVIEWERS USE?

Abstract reviewers use similar scoring criteria to grant reviewers. They want your research methods to be clear and have enough information to evaluate. They expect your conclusions to be justified by your data. They are looking for new methods, new information, and new concepts. They prefer abstracts dealing with important problems that advance scientific knowledge.

COMPONENTS OF AN ABSTRACT

Some organizations require a structured abstract with headings for each section. Even if this is not the case, you should provide a tight and explicit structure for your abstract. This structure focuses your writing and ensures all of the elements are present. It also makes it easier for reviewers to find the specific parts of the abstract. You can announce each section by saying, for example, "The purpose was . . ."

Title

Your title should be concise and unambiguous. It should represent a mini-abstract of your work. It will create the first impression of the abstract for reviewers and, if accepted, for attendees. You do not want the title and authors to occupy an excessive amount of the total word count allotted for your abstract. Your title should be forceful and should begin with an important word. Never (or hardly ever) begin a title with "A" or "The." Your title should include your independent variable, your dependent variable, and the population or species studied. It should be provocative enough to attract an audience. Mark Schuster, MD, PhD, had an abstract title that garnered a lot of attention: "Sexual Practices of Adolescent Virgins." This study showed that adolescents were engaging in behaviors that put them at risk for acquiring sexually transmitted diseases in the absence of what many of them considered to be sexual intercourse. The apparent internal contradiction encouraged the reader to go from the title to the substance of the abstract. You may wish to make a declarative statement in the title, but do not overstate your conclusions.

Authors

The authors should include only those who made substantive contributions to the research. In biomedical research disciplines, the following is the usual order of authorship. The first author has done most of the work. The last author is usually the senior person providing the overall direction to the research project. Helpful colleagues who are not major contributors to the project can be mentioned in the acknowledgments section of your manuscript. Authorship on the abstract is usually the same as that for the manuscript, unless the manuscript combines more than one abstract, new data have been added since the abstract, or the first author on the abstract does not prepare the manuscript. By agreeing to the order and listing of the authors on the abstract, each author is endorsing the authorship specified therein, and it may be difficult to change the order of the manuscript authors.

Agree on the Order of Authorship at the Outset of the Research

Dr. Jones was included as a middle author on an abstract prepared by Sandy in Dr. Smith's group. Sandy was first author and Dr. Smith was the last author. The work was accomplished by Sandy in Dr. Smith's laboratory. Dr. Jones provided an essential reagent for Sandy's work. All the collaborators approved the abstract before it was submitted. When Sandy prepared a manuscript based on the abstract, she used the names and order of authors from the abstract. Dr. Jones objected, claiming that he deserved equal status with Dr. Smith. Dr Jones wanted either first or senior (last) authorship and threatened not to sign the transfer of copyright form required by the journal if he did not get his way. Dr. Smith decided that it was more important for Sandy to be first author and so allowed Dr. Jones to be last author, and Dr. Smith became the subsenior (next to last) author. Dr. Jones agreed to this. Sandy objected, stating that Dr. Smith should be the last author. Dr. Smith explained to Sandy that Sandy deserved the credit for doing the research and preparing the manuscript. Dr. Smith also stated that they would not collaborate with Dr. Jones again.

Introduction

Your introduction should provide the context for your investigations and indicate why you are interested in this area of research. You should provide insight into the foundation for the research problem in one, two, or (rarely) three sentences. This should be written to show that your research was the next obvious step, given the previous state of knowledge.

Purpose

The purpose indicates why this research was undertaken. It provides a statement of your hypothesis in one sentence. It may be considered the specific aim of your work.

Methods

The methods section includes your experimental design. You should specify the animals or population studied, your control group(s), and how they were selected. For standard techniques or treatments, you need not specify the details. If you developed a new technique, you need to provide more specifics. You should include any statistical approaches you used for data analysis. Two or three sentences should be allocated for the methods section.

Results

In the results section, you will summarize your data thoroughly but succinctly. Whenever your data allow, your presentation should be quantitative, not qualitative. Be cautious about using tables in your abstract. If you use a table, you still need to describe the data in prose and probably will not have room to do both. Results can usually be summarized in two or three sentences. However, since this section will sell the abstract to the reviewers, if you can afford an extra one or two sentences in the Results, you should include them. Shorten other sections, if necessary, to include more results.

Discussion

A brief, one-sentence discussion may be included, but only if it adds material not in the Results or Conclusion. Many outstanding abstracts will not have a Discussion section.

Conclusion

Your conclusion should state in one sentence your main point responsive to your purpose. Do not include in your conclusion a promise of more data at the meeting or reach conclusions supported only by preliminary results. Your initial results may not hold up to repetition, or you may not be able to accomplish the research necessary to permit acquisition of the additional data you promise. In either case, you may have to withdraw your abstract, which is embarrassing and will lead others to question your scientific credibility. If you have data in which your are confident, present it, because abstracts are selected on the basis of data, not promises.

Be careful not to phrase a speculation as a conclusion. Your conclusion should be tightly linked to the results. Do not be tempted to extrapolate beyond your data.

Speculation/Recommendation

You can present the implications, importance, and future directions of your research in one sentence. If there is a practical implication for your results, present this as a recommendation describing how your work may alter current practice. Such recommendations must be firmly supported by your data.

SUGGESTIONS TO IMPROVE THE CLARITY OF YOUR ABSTRACT

Give Yourself Adequate Time to Prepare Your Abstract

You should never wait until the day the abstract is due to begin writing. You need to allow adequate time for multiple drafts and for your co-authors to review your near-final draft. E-mail the abstract with the message that if you do not hear from your co-authors by a specified deadline (at least twenty-four hours beforehand), you will assume that they have no changes.

Organize Your Time Effectively for Successful Abstract Preparation

Lee always says he works well under pressure. He likes to wait until the day the abstracts are due to write it so he has as much data as possible. Not only does Lee wait to write the abstracts, but he also postpones reading the instructions until the day they are due. Lee is submitting three abstracts to a national meeting. The first abstract covers research performed by Lee's graduate student. Lee is teaching his graduate student his own bad habits. Lee would like to nominate this graduate student for the student research award, but does not have time to prepare the required letter of nomination. The second abstract involves a co-author in Europe. By the time Lee has this abstract written, his European colleague has left work for the day. She does not see the abstract until the next day, when it has already been submitted. Objecting to the inclusion of very preliminary data that have not been confirmed, she demands that the abstract be withdrawn. The third abstract is poorly written and contains several errors that confuse the reviewers. This abstract is not accepted for presentation.

Your Abstract Should Be Strong and Hard-Hitting

Use key phrases to indicate structure in unstructured abstracts: "the purpose"; "we studied"; "we found"; "we conclude"; "we speculate."

Own your research. Use the first person with the active voice: "we observed" instead of "it was observed that." You will find that it takes fewer words. You should use the present tense for previously published results or for conclusions and speculations that extend beyond the present study. Use the past tense when referring to results of the current study.

Abstracts Must Be Clear and Easy to Interpret

You should avoid abbreviations unless they are standard, and even those should be kept to a minimum. Scientific jargon makes an abstract difficult to comprehend for reviewers or readers who are not in your particular field. The reviewers may have hundreds of abstracts to read and assess in a weekend. Abstracts not clearly written will suffer in the review process.

Read and Follow the Directions

Follow the instructions for submission. Your abstract may be rejected if you fail to do so. You also want to be able to nominate members of your group for the awards for which they are eligible. In addition to the recognition they will receive if they win, there may be assistance with their travel to the meeting. By thinking of those involved in your research, and by demonstrating your appreciation of their hard work, you are also being a good mentor.

Electronic Submissions Require Advanced Planning to Ensure a Successful Submission

Systems for electronic submission are getting better, but they still may be overwhelmed close to the deadline, since many of your colleagues will have procrastinated. While deadlines may be extended under these circumstances, you can spend a lot of time being frustrated while you try again and again to submit your abstract. To meet review and publication schedules, the professional organization sets an arbitrary due date for abstracts. Why don't you set your own date a week ahead of the official due date and avoid the log jam? You also need to be sure that you receive a confirmation receipt.

AVOID BAD ABSTRACTS

You want to avoid the problems of bad abstracts, which will have a negative impact on your reputation. You do not want your abstract to be unreadable due to bad writing, abbreviations, or jargon. You do not want your abstract to lack substance because you have too few subjects (avoid case reports that contribute nothing new), your data are too preliminary, or you just make promises. Promises may not be fulfilled, causing you to withdraw the abstract and be embarrassed.

BUILD YOUR CAREER ON QUALITY

You want to be sure that your abstract is new, important, and exciting. Even if you have your heart set on presenting your work at a particular meeting, if it is not ready, you should wait until it is. Take the long view of your career and emphasize quality. It takes years

Table 3. *Timeline for Abstract Preparation*

	Time before deadline					
	2 yrs.–3 mos.	*2 mos.*	*1 mo.*	*2 wks.*	*1.5 wks.*	*1 wk.*
Select a research topic	✓					
Complete research	✓	✓	✓			
Prepare abstract			✓			
Send abstract to coauthors				✓		
Revise abstract					✓	
Submit abstract						✓

to build a reputation for strong and careful science, but it is easy to destroy such a reputation with an abstract that has to be withdrawn or one that is lacking in substance.

Replication Ensures a Quality Abstract

Kelly was working on a new project. It took longer than expected to determine the research parameters and collect the initial data. When the abstract deadline approached, the experiment had worked once. Kelly was disappointed when his mentor suggested they wait and replicate the experiment before submitting an abstract. This meant that Kelly did not present at his original target meeting, but when he was able to replicate his results, he had a better abstract that was selected for a platform presentation.

CHAPTER 4

Presenting at
Professional Meetings

PREPARE A PROFESSIONAL,
POLISHED PRESENTATION

Congratulations, your abstract was accepted for presentation at your meeting. Remember the amount of effort and funds you expended in producing the results summarized in the abstract. At this point you should not scrimp on the cost of materials or on the effort required for an effective presentation. We have all witnessed less-than-professional presentations by established colleagues as well as by beginning trainees. A little planning and preparation can lead to a polished presentation.

EFFECTIVE POSTER PRESENTATION

Prepare a Poster with a Professional Appearance

Rely on your abstract to form the basis of your poster. There is no need to re-create the content of your poster presentation. Your abstract serves as an outline for your poster with the addition of fig-ures, graphs, and tables. The title, authors, and affiliations should run across the top with any logos. The letters should be at least one inch tall. The rest of the poster should be organized in columns with headings in the following order: abstract, introduction, purpose, methods, results, conclusions, and speculation/recommendation. Use color to attract attention, but do not overdo it. Each figure, graph, or table should have its own legend. The print for the body of the poster should be readable at a distance of at least three feet.

You can print the whole poster on a single large sheet and lami-nate it to protect it when you practice with your group and during transport. You will need a tube to transport and protect the rolled-up poster. Before printing your poster, you need to know the size and orientation of the display area for the poster. When the poster does not fit the area assigned to it, the appearance detracts from the substance and it may be quite awkward, particularly if it overlaps onto your neighbors' posters.

A less expensive format that can be easily transported in a brief-case or suitcase is as follows. Your poster can be printed out as the title, authors, and affiliation on a strip, and each section of the poster being on individual 8½ by 11 inch paper. These can then be lami-nated to protect them during practice and transport.

Remember That Your Poster Is Not Comprehensive

Your poster should be used as an outline for more in-depth discus-sion with those who come to view it. In addition to a well-written abstract, visitors will be attracted by your creative use of color and your overall visual presentation. You might consider color schemes similar to those in your PowerPoint presentation. Data will also attract attention, so be sure to display your results in a logical and appealing manner.

Your Poster Provides an Opportunity to Interact with Colleagues

You should put up your poster and remove it in compliance with the instructions of the meeting organizers. You should be present during the entire attended viewing time for your poster as described in the meeting program. If you are presenting more than one poster at the same time, one of your colleagues should be with the other poster(s) at the same time. If that is not possible, then leave a note on the unattended poster(s) giving the poster number where you can be found for discussion. Be sure to bring your business cards to distribute to contacts you make. These are especially important if you wish to receive reagents or clinical material from contacts you make at the meeting. You might also consider printing a copy of your poster to distribute and bringing copies of your most recent, relevant publications to give to interested parties. Remember that presentations, poster and platform, provide opportunities to market yourself and your work.

In the past, posters were considered less desirable than platform presentations. However, posters provide an opportunity to meet colleagues in your field and to discuss your work with them. Often, very senior scientists, who you may not feel comfortable approaching, will come to your poster. Mentors should attend the posters of their trainees or junior faculty members in order to introduce them to colleagues. However, the junior person should answer questions and engage in any discussions, with the mentor only providing support to the junior person and helping out if the crowd around the poster becomes too large for the junior person to handle.

One colleague of ours talks about meetings in general and posters in particular as providing "face time." Present your best work effectively and use your talks and posters as opportunities to network.

EFFECTIVE POWERPOINT PREPARATION

If you've been invited to give a talk, in most disciplines, you will be giving a PowerPoint presentation. Some disciplines do not use PowerPoint, in which case you should follow the convention of the meeting. For example, some venues want a "chalk talk" for more spontaneity. Be sure to practice the chalk talk numerous times so that you are comfortable with, and confident in, your presentation.

Follow the "Rule of Sevens" to Produce Effective Slides

The most common errors in PowerPoint presentations are too much writing, too much information, and too small a font. The "Rule of Sevens" specifies that no PowerPoint slide should have more than seven lines and no line should have more than seven words. There is a temptation to use complicated tables or figures that have already been published. In a large venue, your audience will not be able to read the print and will be distracted by extraneous information. You should take the time to develop simpler tables and figures that specifically make the point you want to make in your talk.

Large Rooms Require Even Greater Care

If you will be talking to more than a thousand people, you need to increase the size of your font and further simplify your PowerPoint slides. You do not want your audience distracted while they strain to read a font that is too small. Not only will they not be able to read what is on the screen, but they also will not be listening to what you are saying if they are focused on deciphering your slides.

You Should Use the Same Format throughout Your Presentation

You should select a specific format for your PowerPoint presentation. If you need to combine material from several talks that are in different formats, you should choose one format so that it is a uniform, seamless presentation. The title for each slide should be in a larger font than the text. The color scheme should make the lettering easy to read, such as with dark letters on a light background or vice versa. Do not use red letters or red line drawings; they always look great on your computer monitor, but they do not project well. If you shade the red to pink or orange, you might be able to use that color for emphasis. You should be cautious and project your presentation so that the audience will be able to see all of it. Use shading to give depth to your letters. If you want to add the logo of your department and/or institution, place it in a corner of the slide and be consistent with its placement, using it as a part of your slide template for the entire presentation.

Your text should consist of bullets with phrases. This should be an outline, not a paragraph. Your talk should add information beyond what is printed on the slide. You do not want to simply read the slide, but embellish it to make it more interesting to your audience.

Your Abstract Forms the Outline of Your Presentation

Your abstract already contains the structure for your ten-minute talk. Use your abstract as an outline and expand with additional content or new information. A ten-minute talk will usually include fifteen to twenty slides, if you follow the Rule of Sevens and do not put too much information on your slides. You should practice ahead of time to be sure your talk fits the allotted time.

Avoid Putting Too Much Information in Your Tables and Figures

We have all seen slides with so much information that they were incomprehensible. Tables should contain no more than two rows by four columns, or three rows by three columns. If it takes you longer than forty-five to sixty seconds to explain, it is probably too complex and you need to simplify, perhaps by presenting the information on two slides. Tables from journal articles have too much information and should not be used. Some presenters will tell the audience to ignore rows and columns from published tables, but it is better to summarize the important information in a new table. Bar graphs should have no more than six bars. It is important to label the axes. Pie graphs should include numeric percentages for each part of the pie. Most line graphs should have a zero origin for both axes, the axes should be labeled, and rarely should the equation for the line be a part of the graph.

Know the Rules and Conventions of the Organization

For example, one national organization banned acknowledgment of the presenter's collaborators. The rationale was that this was taking time away from the science, and was often redundant since the coauthors were named on the abstract. Collaborators not named could be acknowledged in the manuscript. There are also cultures specific to different organizations. Use your network of colleagues to learn the culture. Ask before you leave home, so you can practice your talk in accord with the expectations of the cultural setting in which you will present.

THE TEN-MINUTE TALK

Your Goal Is to Make Sure Your Audience Remembers One Thing

Your goal in a ten-minute talk is to leave your audience remembering one important take-home lesson from your presentation. Only your closest colleagues and fiercest competitors will remember more detail, so focus on making your key point effectively. Remember the adage "Tell them what you are going to tell them, tell them, and tell them what you told them."

Your PowerPoint Slides Organize Your Talk

In preparing your talk, you should first make your slides and then practice. You should write your talk to explain your slides. There are two very different points of view regarding reading versus talking in your presentation. If your discipline favors one over the other, you should follow that convention. If your discipline is neutral on this issue, you should decide which format is most comfortable for you. Regardless of the format you choose, try to have your talk ready two weeks before your meeting so you can practice the talk with your group and can revise it as suggested by your own review and that of your colleagues.

A Ten-Minute Talk Should Actually Be Ten Minutes Long

A ten-minute talk is very difficult because you have so little time. The organization must be tight without overwhelming the audience with too rapid a delivery or an excessive amount of detail. We all have seen individuals who give the same presentation as a ten-minute talk or a one-hour talk, simply faster or slower depending on the time allotted. They also present the same information regardless of the audience, ranging from the public to students to colleagues. Practice in front of colleagues will help establish your rate of presentation and help you determine whether or not you have the right number and right type of PowerPoint slides. Remember, in general, fifteen minutes have been allocated for your presentation: ten minutes for the presentation and five minutes for questions. If it takes you only five minutes to present your talk, you will have ten minutes for questions.

Always confirm the time allotted, because the format may be different depending on the venue within the meeting. For example, platform presentations may be longer if selected for the plenary session—perhaps a total of thirty minutes instead of fifteen. Check to be sure that there will be time for questions, because some plenary sessions do not include time for questions.

Practice to Determine How Long Your Talk Will Take

Kim was preparing to give his first talk at a meeting. In his initial practice presentation at lab meeting, it took two and a half minutes out of the fifteen allotted. Kim's mentor quipped, "Well, you will now have to answer questions for twelve and a half minutes." The next time Kim practiced, his talk lasted for close to ten minutes.

PRACTICE FOR THE QUESTIONS AFTER YOU PRACTICE YOUR TALK

The most important aspect of practice is the questioning by your colleagues. Too many presentations are derailed when the presenter is unable to answer the simplest question. You need to have colleagues and mentors who will ask you the difficult questions at home, and help you to formulate your answers. If you anticipate an especially aggressive rival in the audience, one of your local colleagues should assume the rival's persona and challenge you with the questions you expect from such a rival. Practice answering questions succinctly in sound bytes of no more than three sentences. If the answer is not thorough enough, the questioner will follow up with another question. Answers that are too long will cause the audience to lose interest. If you get unexpected questions at the actual presentation, your practice sessions were not sufficiently thorough.

Remember that you do not have a contract with each of the members of the audience. If someone asks a particularly difficult question, you can suggest that it might be better to discuss this privately after the session. If someone is very aggressive with a question, you can say, "Thank you," and go on to the next question. The audience will thank you silently for not getting into a verbal fight with the aggressor.

Bad Answers Will Be Remembered Even If the Talk Is Good

Lynn is a graduate student preparing to give his first ten-minute talk at a national professional meeting. Lynn's mentor assists him in preparing his PowerPoint slides and organizing his talk. The department chair always organizes a practice session for those selected to give presentations at the meeting. Lynn practices his talk at this session, and it goes very well. The audience discusses Lynn's presentation in general, and then Lynn goes back through his talk one PowerPoint slide at a time with the audience making appropriate suggestions for changes. With the feedback Lynn receives, his presentation is greatly improved. Lynn makes a flawless presentation at the meeting. However, when the questions begin, Lynn realizes he did not truly understand his project. The audience soon realizes this as well when Lynn stumbles over his answers. Some of the answers take more than one minute, during which Lynn talks around the topic but never answers the question. Other answers are only a single word, "Yes" or "No."

Answering Questions Is a Learned Art

Lee's mentor understands the importance of practicing not only the talk but also the questions. Lee practices in front of her lab group. In addition to providing feedback on Lee's talk, her group asks more than ten questions. Some of these are very difficult, so Lee has the opportunity to answer them and receive feedback. She also receives suggestions regarding her style of responding to questions. Lee is taught to briefly restate the question before answering, in case not everyone in the audience is able to hear the question. This also gives her more time to consider her answers, which is especially helpful if the question is difficult. Part of Lee's training includes the admonition that it is appropriate to say that she does not understand the question and to ask for clarification from the questioner or the chair of the session. In addition, she can admit to not knowing the answer if that is the case, or suggest that they discuss this with her mentor after the session. She knows to keep her answers brief, no more than two or three sentences. If asked about work in progress, Lee will be ready to respond, "This is an interesting question. You have anticipated the course we are taking on this project, and we are currently pursuing that line of research."

THE NUMBER OF POWERPOINT SLIDES FOR YOUR PRESENTATION SHOULD BE PLANNED AHEAD OF TIME

A general rule is one and a half slides per minute, or about fifteen slides for a ten-minute talk. If the slide is a simple word slide or photograph, you can expect to spend thirty seconds on it. If the slide illustrates a complex model or procedure, plan to spend more time, but rarely more than one minute. If a slide requires more than one minute, it is probably too complex for a ten-minute talk, and you should try to simplify the information on the slide or break it down into two slides.

Your talk is essentially an expansion of your abstract. If permitted, you should have one slide with title, authors, and affiliations. You can use a color version of your data or model or some other unifying figure as a background for this slide, if it is appropriate and not too busy. If the session moderator reads the title and the names of the authors when the speaker is introduced, there is no need for you to repeat this. There should be one or two slides for the Introduction. The Purpose should take one slide. Between one and three slides can be used for the Methods. Three to five slides are usually needed to present the Results. If you use Discussion or Summary slides, there should be only one or two. Conclusions can be covered in one or two slides. If you have any Speculations or Recommendations, use one slide for each. Ending with a slide of all of your collaborators is an unnecessary use of precious time. If these individuals are co-authors on the talk, they have already been acknowledged in the abstract, on the first slide, and/or in your introduction by the chair of the session. Far better for you to end your presentation by simply saying "Thank you" to let the audience know you are done; they can applaud, and now you can answer questions.

BE SURE TO ACKNOWLEDGE
THE WORK OF OTHERS

How would you feel if a speaker was showing a slide containing your data without indicating that this was your work? If you are using someone else's data, you should provide a citation as part of the title or at the bottom of the slide and thereby acknowledge the source in your talk, naming the senior author of the group (e.g., "from Dr. Smith's group"). You do not want to give the impression that you are taking credit for work done by others.

PREVIEW YOUR PRESENTATION AND THE PODIUM,
AND BE GRACIOUS BEFORE, DURING AND
AFTER YOUR PRESENTATION

Be sure to allow yourself time to load your PowerPoint presentation in the Speaker Ready Room. Most professional meetings specify that you should do this the day before your talk. If you have video as a part of your presentation, be sure to upload the necessary supporting software files linked to your PowerPoint presentation. Use the computer in the Speaker Ready Room to practice your presentation and to be sure everything is there and any video is working.

Go to the room where you will make your presentation before the session begins. Introduce yourself to the moderator and be sure they know how to pronounce your name, the names of your co-authors, your affiliation, and the title of your presentation. Introduce yourself to the computer technician and confirm that your presentation is available. You can also ask them to quickly run through your presentation to be sure it is all there and any video is working. You should check out the podium, and move it if necessary so you don't fall off the stage during your presentation (this has happened). You need to be sure that you understand how to adjust the microphone, to advance your slides, and to work the pointer. If the stage is placed so that you cannot use the laser pointer, then practice using the curser as the pointer.

During your presentation, speak slowly. Most of us tend to talk faster when we are nervous. If there are problems with your PowerPoint slides, microphone, or pointer during your presentation, do not lose your composure or your temper. Do not get angry with the computer technician, since that individual determines your destiny during that fifteen-minute period. Your audience will remember you if you handle an unfortunate situation with grace, and they will also remember if you do not.

If you have a questioner who is arrogant, aggressive, or naïve, handle yourself with poise. The audience will be on your side, and will respect you if you respond evenly and effectively. If the questioner is particularly obnoxious, you have no need to respond—you have not entered into a social contract with them that demands a response. As we noted previously, just thank them and move on to the next questioner.

If possible, stay for the entire session. After all, you wanted an audience for your presentation, so be there for the other presenters.

Before you leave, be sure to thank the moderator and the computer technician. You may find yourself working with them again in the future. For some of the larger meetings, the same technicians are there year after year.

Making the Best of a Bad Situation

Dr. Jones was the third and final speaker during the first session of the annual meeting. The venue was one quarter of a large ballroom, and there were sessions in each of the other three quarters of the space. As the moderators began introducing the first speakers, all moderators could be heard in all four sessions. The computer technicians turned off the microphones while they worked on the problem. The sessions went on, and the microphones were working properly by the middle of the second presentation. Dr. Jones was the first to use PowerPoint in his session. Unfortunately the projector was not working. Dr. Jones just kept talking. Luckily he did not have any tables, graphs, or drawings essential to his presentation. The projector began working in the middle of his presentation. He just kept going. He received lots of accolades after his talk for his professionalism. It was good that he had printed out his PowerPoint slides and had the hard copy with him in case of a malfunction.

CHAPTER 5

Marketing Your Ideas through Publications

Your abstract has to be focused on one message, but a whole publication can develop more than a simple message. Therefore, you may use one or several abstracts as outlines for your publication, or often you will develop the outline of the publication de novo, relying on the data from abstracts but recognizing the need for a different organization. The important point for you is that you are marketing your logic, ideas, and professionalism through your publications. These are the currency with which you will build your career and your funding.

FOCUS YOUR RESEARCH PUBLICATIONS

If you focus your research articles on a specific topic, you will find that before long you are the go-to expert in this area. Let's say that you are a relatively new faculty member with several years of experience and you have eight first or senior-author publications. If these are all narrowly focused on a very specific topic, you will be considered an expert on that topic. You will then have credibility if you suggest a review on that topic; will be asked to be a reviewer for manuscripts in the area; and may be asked to write a commentary to accompany a manuscript you reviewed (see below). If, on the other hand, these eight publications are on several different topics, you will be considered unfocused and will not have any of these opportunities.

Select a Journal

Before you select a specific journal, you need to consider the type of journal. Do you want a basic or applied journal? Are you concerned with sharing fundamental information, or do you want to impact practice? Do you want a general or a specialty journal? Would you like to have a broader impact, beyond the limits of your subspecialty, or does your work have a specific focus within a particular subspecialty? Do you want to publish in a new journal or in an established one? A new journal may have a quicker turnaround time—less time between submission and a decision. Some new journals, in contrast, do not publish in a timely fashion, and there is a risk that they may not survive. In addition, a new journal may not be listed in some of the Internet citation search engines, such as PubMed, for at least the first two years of publication. If other authors can't find your work, they can't cite it.

Pick Three Journals

Once you decide on a specific type of journal, you need to determine the level of prestige in which you are interested. You should select three journals at three different levels of prestige and write this list down in a place where you will be able to find it. First, submit your manuscript to a journal that you feel is extremely competitive and may not accept your work. You should feel you have about a 5 percent chance of success with this journal. Your second choice should be one you are reasonably sure will accept your manuscript. You should feel you have about a 50 percent chance of success with this journal. In case this journal does not accept your manuscript, you should have a third journal with which you know acceptance is nearly guaranteed. You should feel you have about a 95 percent chance of success with this journal.

Having this list of three journals before your initial submission will help protect your work from delayed resubmission if your manuscript is rejected by your first, and even your second, choice. We know that when we are rejected by a journal, we tend to take it very personally. But by having this list ahead of the first submission, we are anticipating possible rejection. We can move forward more quickly if rejected, and if we are accepted by one of our first two choices, we have done well. To prepare yourself, you should download the Instructions for Authors for each of the three journals so you are prepared to make any structural changes to your manuscript to conform to the standards of the next journal. This process also helps you to try for the highly competitive, prestigious journals, because if you do not try, recognizing the substantial risk of rejection, your work will not be published there. Also you are recognizing that manuscript review is capricious, and rejection does not reflect personally on you. Knowing this before you submit your manuscript will help you deal with a rejected manuscript. You will be in a position to make any changes requested by the reviewers and any modifications required by the Information for Authors before you submit your manuscript to your next choice. You should submit your revised manuscript within one week of rejection by the previous journal, unless the reviewers' comments legitimately identify the need for additional data.

In addition to weighing prestige, you need to consider the length of time from submission to appearance online and from appearance online to publication. The most important interval is that between submission and appearance online. You want your work to be available so that others may read and cite it. You may be able to negotiate with the editor of the journal for an expedited review, even if the journal does not have this policy.

A journal's impact factor is something you should consider. Published every year, this score indicates the average number of citations for each article published in the journal. The length of articles is also included in the score, which is why some journals limit the number of words per article. Some institutions, notably those in Europe, are very concerned with impact factor for promotion and tenure. If you look at the journals with the highest impact factors, you will find that the most frequent number of citations for a single article is one. The same is true for journals with the lowest impact factors. The difference in the average impact factor is due to the few articles that are cited very frequently in high–impact factor journals. Especially in the United States, promotion and tenure committees are beginning to ask faculty members to indicate their most important work and to consider the impact factor of these particular articles, not an average of the articles of other individuals who also publish in these journals. They are also considering the number of patents, in addition to the number of grant dollars.

Know the Current Editorial Policies for the Journals You Select

When you review the Instructions for Authors, pay special attention to editorial policy. Can you submit your manuscript to a specific editor who knows your work? Are authors able to recommend reviewers? You cannot recommend those with whom you have personal or professional relationships (e.g., family members, mentors, current collaborators, individuals with who you have joint publications within the last five years, or colleagues at your institution), but you should suggest leaders in your field and those who know your work. Even if the journal does not ask you to recommend reviewers, you can do so. You should recognize that the editor may or may not accept your recommendations. You can always specify that your mortal enemy should not be allowed to review your manuscript. You do not have to give a reason. Manuscripts have been delayed or rejected by editors or reviewers who are competitors or friends of competitors.

You Cannot Exclude the Whole Field from Reviewing Your Manuscript

Dr. Johnson worked in a very competitive field. There were basically two camps, and you were in either one camp or the other. She submitted a manuscript with co-authors who were members of her camp. She asked the editor to exclude each member of the other camp from review. The editor faced an interesting dilemma. Would there be enough individuals to review given the number of co-authors and the number of excluded reviewers? Fortunately, the first two potential reviewers agreed to review the manuscript. There were only three individuals whom the editor believed to have the expertise to review this manuscript and who were not co-authors or exclusions.

If you cannot name reviewers or specify a particular editor, you should consider selecting a journal based on the members of the editorial board. Members of the editorial board will probably determine reviewers or be reviewers themselves.

You should seek the advice of senior faculty in your field regarding your selection of journals. Scan recent issues of the journal to see whether it publishes articles in the format in which you wish to submit. For example, if you are considering a review article, even if a journal's editorial policy states that it accepts review articles, if none have been published in the past six months, you should not submit one. If the journal has recently published an article on your topic, that suggests that its board may be interested, but, on the other hand, some journals do not want too many articles on the same topic.

Request Exclusion of Your Nemesis as a Reviewer

Lee and Lynn each prepared a manuscript at approximately the same time, and they submitted their manuscripts to the same journal. Dr. Smith, a mutual nemesis, was a member of the editorial board. Lee requested that Dr. Smith not be allowed to review his manuscript. Lynn did not. One month later, Lee was asked to make some reasonable changes in his manuscript and submit a revised manuscript, which was accepted for publication. Lynn still had not heard from the editor of the journal. Lynn spoke to the editor, who said she was having a hard time completing the review process. Lynn reminded the editor that his manuscript had already been under review for three months. When the editor would not make a specific commitment to Lynn regarding his manuscript, Lynn withdrew his manuscript from consideration and immediately submitted his manuscript to another journal. Lynn still did not ask that Dr. Smith be excluded as a reviewer. Lynn's manuscript was rejected by the second journal. Lynn's manuscript was finally accepted by the third journal, a full year after Lee's manuscript had been accepted by the original journal.

Do Not Violate a Prepublication Embargo

Some journals have a prepublication embargo. There cannot be any prepublication publicity about your work if you want your manuscript published in the journal. These journals typically work with the author's institution to attract attention to the paper at the time it is available online.

Flattery May Tempt You to Violate an Embargo

Kelly submitted an abstract to a national meeting. The abstract was about a very exciting finding. The meeting organizers selected Kelly's abstract as one of the ten best at the meeting. Kelly was invited to meet with members of the press to discuss her research. Kelly was flattered by the invitation, but had to decline. She had already submitted a manuscript on this work to a journal that had a prepublication embargo. Meeting with members of the press before the article was online would prevent its publication in this journal.

Follow the Instructions for Authors

Visit the journal website to obtain the Instructions for Authors. You need to follow the general and reference formats. Some editorial offices will return manuscripts that do not conform to the instructions, causing unnecessary delays. Also, the editor may suspect that you submitted the article elsewhere and were not careful to revise the manuscript for the second journal. Remember, your manuscript is a reflection of the care you take with your research work and helps establish your reputation with the editor. If you cannot follow the well-described Instructions for Authors, the editors and the reviewers may be concerned about your ability to follow research protocols correctly and to report results accurately.

Remember to Include an Effective Letter of Transmittal

Your letter of transmittal is an opportunity to sell your manuscript to the editor. It should be no more than one page in length. You should state the title in the first sentence. You should briefly describe the background to your work, your methods, your results, and your conclusions. Place your work in context and elaborate on any important breakthroughs and implications. You should cite the relationship between your manuscript and any work recently published in that journal or elsewhere in the literature.

If the journal requests that you suggest an editor or reviewers, do so. Request that your nemesis be excluded from reviewing. You should include any required statement or form regarding copyright.

You should specify that this work has not been published previously and is not currently under review with another journal.

You Can Publish Results as Original Data Only Once

Dr. Jones was a new assistant professor who was feeling a lot of pressure to publish. Dr. Jones decided to publish original data in a chapter for an edited book. He also wanted to submit the same data as original manuscripts to two journals simultaneously, and he would publish in the journal that accepted his manuscript first. Dr. Davis was asked to review Dr. Jones's manuscripts by the editors of the two journals. Dr. Davis immediately contacted the editors and suggested that Dr. Jones had submitted the same manuscript to both journals at the same time. Dr. Davis also mentioned that Dr. Jones had already submitted these data for publication to the book he was editing. The editors immediately rejected Dr. Jones's manuscript and warned him to never again submit previously published material or simultaneously submit the same manuscript to two different journals. In a rush to publish, Dr. Jones has seriously damaged his professional reputation with the journal editors and Dr. Davis. In addition, he has traded a chapter in an edited book for an original first-authored manuscript, the latter being much more valuable to Dr. Jones's promotion and tenure.

Remember, in most cases the publisher owns the copyright on your published work. You do not. Therefore, you cannot publish original data more than once, unless you reference the original publication and note that those data were published in that report.

You should close your letter of transmittal with a statement thanking the editor for a rapid review, so you end on a positive note. You should expect a return e-mail acknowledging receipt of your manuscript and including a manuscript number. Use this manuscript number in any future contact with the editorial office regarding your manuscript.

You Can *Plagiarize Yourself*

Dr. Johnson received many invitations to speak at international confer-
ences. She accepted as many invitations as she could in order to build her
international reputation. Most of these conferences requested that she sub-
mit a written version of her presentation for publication in the confer-
ence proceedings. Since the talk she gave in Barcelona was essentially the
same as the talk she gave in Tokyo, she submitted the identical manu-
script to both meetings. When the publisher of the Barcelona proceed-
ings saw that Dr. Johnson had an identical paper in the proceedings of
the Tokyo meeting, the publisher banned Dr. Johnson from publishing
in its journals and publicly announced this ban in each of those journals.

You need to provide a different manuscript for each publication so
you don't plagiarize yourself. In addition, we discourage submission
of any original data in conference proceedings. Conference proceedings are frequently not readily available, and others cannot reference them. We know of individuals who have been sanctioned by
journals for duplicate publishing in a conference proceeding and as
an original report in a journal.

Maintain Communication with Your Co-authors

Prior to submission, your co-authors need to receive the manuscript
within a reasonable time for them to provide their comments.
Indicate that if you have not heard from them by the deadline, you
will assume they have no suggestions for changes. When you submit the manuscript, e-mail them a final copy. If you need to revise
and resubmit the manuscript, you should send them copies of the
reviewers' comments and the letter from the editor along with your
revised manuscript for their review. E-mail them when the manuscript has been accepted. Just like you, they need to know the status
of the manuscript for their curricula vitae, promotion, and tenure

and for citations in their own manuscripts. It is very frustrating to discover that you are a co-author for a published paper of which you were unaware. To monitor the status of your accepted manuscripts and to uncover any that you were not aware of, you should do an electronic search on a regular basis, such as weekly or monthly.

Your Co-authors Need to See Each Version of the Manuscript

Chris was the corresponding author of a manuscript with ten co-authors. He requested input from his co-authors for the original version of the manuscript. When he was asked to revise the manuscript, he simply made the changes requested and submitted a revised manuscript. When the revised manuscript appeared online, several co-authors contacted the journal editor and asked that the manuscript be withdrawn. They had problems with the revised manuscript and were concerned that they had not been involved in the revision. The editor rejected the manuscript and suggested that Chris revise the manuscript through a process of consultation with his co-authors. The editor would not consider the new submission until each co-author sent an e-mail with electronic signature confirming that they had reviewed and approved the new submission. Chris placed the editor in a very difficult position in a rush to revise and resubmit.

Journals Lose Manuscripts

A very prestigious journal was producing a special issue devoted to Sam's discipline. Sam submitted a manuscript for the special issue. After the reviews were completed, the editor said that Sam's manuscript was of interest, but not significant enough to warrant publication in the prestigious journal. The editor offered to submit Sam's manuscript to a journal franchise title under the same publishing company's umbrella of titles. Sam felt that her manuscript would more likely be accepted for publication and would be published more quickly if she allowed the parent journal to forward her manuscript to the related journal. When Sam did not receive an e-mail from the related journal within a week, she contacted the editor of the related journal and found that the editor had not received her manuscript. The editor apologized for the delay and expedited the review of Sam's manuscript.

After Submission, Keep Track of the Timeline for the Process

Once the manuscript has been submitted online, you should receive an acknowledgment from the editorial office within twenty-four hours. If you don't, you should contact the editorial office to be sure that it received the manuscript. If you don't hear from the editor within two months of submission, contact the editorial office to determine the status of the manuscript.

Do Not Attempt to Micromanage the Review Process

Lee was anxious to have his manuscript published. One week after he submitted it, he e-mailed the editor to ask when he could expect a decision. The editor replied that two individuals had agreed to review the manuscript and their reviews were due in less than two weeks. Lee was concerned that this was not fast enough. He contacted the journal representative in the publisher's office, who referred Lee to the editor. Lee could see online that something had been done with his manuscript two weeks after submission. Lee phoned the editor, asking if a decision had been made regarding his manuscript. The editor explained that one review was complete and the second review was expected in several days. Again, Lee contacted the journal representative in the publisher's office, who referred him to the editor. Lee was attempting to pit the journal representative and the editor against each other. This was foolish on Lee's part. The journal representative and the editor had a strong working relationship forged over years of working together. They resented Lee's attempts at micromanaging. When the reviews were mixed, the editor rejected Lee's manuscript and made a mental note to set a high bar for the original editorial office decision whether to reject immediately or send the manuscript out for review if Lee should make future submissions.

Be a Willing and Conscientious Reviewer

Remember, it is difficult to secure reviewers for manuscripts. When you are asked to review a manuscript in your area, you should do so promptly and thoroughly. Some have asked whether it is a conflict of interest to review a manuscript by a competitor. You should review this manuscript fairly, preferably on the day you are asked to do so. Providing an unfair review or delaying your review would be a conflict of interest.

Reviewing manuscripts helps you to stay up-to-date in your field. It gives you insight into your competitor's research. It can increase the citation of your work if the author has not cited your work and should have. If you write a good review, the editor might ask you to prepare a commentary on the manuscript. The commentary would be another publication for you and is essentially already written if you have done a conscientious review. Reviewing manuscripts can also help you find papers for which you should be a co-author.

You May Find You Are Reviewing Your Own Work

Chris was asked to review a manuscript. When she read it, she found that the authors included some data that she had presented as a poster at a recent national meeting. Chris was furious. There was no reference to her poster. Her initial impulse was to call the corresponding author and demand to know why they had used her research inappropriately. Fortunately, she spoke to her mentor first. Her mentor insisted that this was a matter to be decided by the editor of the journal. The editor asked to see Chris's data and agreed that they were similar to those in the manuscript. The editor asked the corresponding author for the source of the data. The author said that they came from a colleague who was cited in the acknowledgments. The colleague was collaborating with Chris and had provided her data for the paper. The editor rejected the manuscript without review and warned the authors to be sure of the source of every piece of the data in their manuscript. The editor invited Chris to submit a manuscript on her data and suggested that in the future she plan to submit her manuscript before presenting her data at a meeting.

Take the Time to Craft a Good Title

The title of your paper is your initial opportunity to attract potential readers. This is the first thing readers see when they do literature searches, read other papers citing yours, scan the table of contents of a journal, and read your curriculum vitae, grant applications, progress reports, and grant renewals. The title should concisely state the manuscript's content. There should be enough information in the title so the reader can determine their interest in the article, including the essence of the experimental work, the species investigated, and any qualifying phrases to clarify the nature of the investigation (for example, "in vitro"). A declarative title will attract attention but should not overstate your conclusions. For example, if the work was done in cell culture, make this clear and do not imply an effect in the live organism. We spend a considerable amount of time developing the title because it is so important for current and future interest in our work.

One way to develop a title is to craft a series of statements that declare the key message of your work, while you mix and match terms and phrases from your draft titles. Another method is to list the key terms that you would consider for the title and then select from the list to develop a series of draft titles, from which you choose the best. The process of developing a title may stretch over more than one day as you propose and sleep on different possibilities.

Consistency in Your Name Is Important
to Establish Your Reputation

You need to decide on the name you will use in your publications. Once you decide, you should stick with this name. You may want to include your middle initial(s) and a second last name to differentiate yourself from individuals with similar names. If you use a middle name routinely as others would use a first name, then use your first initial and spell out your middle name on your publications. You want readers to recognize who you are.

Authorship Requires Substantive Contribution

Deciding who should be included as an author, and in what order, is often a point of contention in the preparation of a manuscript. Each author should have made a substantial contribution to the manuscript, not just have performed menial tasks as part of their job. Technicians should be included if they have contributed original thought or extraordinary effort. Physicians providing clinical material should be included if they provided clinical insight in diagnosing the patient and an intellectual contribution to the research.

Your boss, the clinic director, or the laboratory director should not be automatic co-authors if their contributions do not extend beyond their administrative roles. Without substantive involvement in the research, being the boss is not sufficient for authorship. To obtain independent grant support for your research, you need to have publications without your boss or mentor to demonstrate your actual independence.

The acknowledgements section can be used to thank these individuals instead of making them co-authors. The acknowledgments should also identify grant support relevant to the performance of this research. You will be asked to submit publications as part of your annual report and your competing renewal for your grant, and they need to acknowledge the grant to be considered relevant to it. If you did not think the funding was important enough to include, why should the grant reviewers? You should also reveal in the acknowledgements any affiliation or support from industry, especially if a conflict of interest might be perceived. A number of senior investigators have gotten themselves in trouble by not acknowledging industry support and having a journalist find out about this later.

Know the Rules for Authorship in Your Group

Kelly is a graduate student working with Dr. Smith's group. The group includes a postdoctoral fellow and two other graduate students. Kelly's research is going really well. After six months, Dr. Smith suggests that Kelly write a manuscript. Dr. Smith indicates that the authors should be in the following order: postdoctoral fellow, Kelly, and Dr. Smith. Kelly asked Dr. Smith why the postdoctoral fellow was even being considered for authorship, since the fellow did not contribute in any way to her project. Dr. Smith replied that the postdoctoral fellow was generally responsible for the research in the group and always received first authorship. Kelly checked with the other graduate students in the group. They reported that Dr. Smith insisted that the fellow be the first author on all of their papers, with Dr. Smith being the last author. Kelly went to the department chair, who agreed that Kelly should be the first author. With the intervention of the department chair, Kelly was made first author and the postdoctoral fellow was not included as an author. However, both Dr. Smith and the fellow were so upset with the department chair's involvement that Kelly had to find another group with which to work.

Order of Authorship Should Be Rational and Meaningful

The order of the authors can be a contentious matter. The first author should be the person performing the research and preparing the initial draft of the manuscript. In situations where the person doing the bulk of the research work and the individual authoring the first draft are not the same, a decision must be made regarding the relative weight of work and intellectual contribution for these different activities. The last author should be the mentor. The order of the other authors should be determined by their levels of contribution. Some papers indicate co-first authorship for the first two authors, but it must be recognized that this information is lost once the paper is cited or indexed. It might be more meaningful for the two individuals to alternate first authorship if their contributions are truly equal and there will be more than one publication in journals with equivalent impact factors. Another option is for the mentor to give up last authorship and one person (perhaps the postdoctoral fellow) be first author and the other be last author.

Mentors Should Cede First or Last Authorship on Collaborative Papers

Kim is a fellow in a high-powered group. Kim's project involves a collaboration between two groups. When it comes time to publish, Kim is dismayed to find the PIs of the two groups in the first and last authorship positions. Kim was second author. The question is how many first- and last-authored papers are enough for the mentor? One hundred? Two hundred? First or last authorship is much more important for a postdoc. Kim's mentor should have negotiated to be subsenior author with the other PI as senior author to obtain first authorship for Kim.

Funding and Previous Publications Are Not Enough to Justify Senior Authorship

Dr. Smith is the chief of the division and the principal investigator on the postdoctoral training grant for the division. Lee is a postdoctoral fellow in the division who is supported by the training grant. Lee's project with Dr. Johnson follows up on some of Dr. Smith's publica-

tions, but shows that the model developed by Dr. Smith is incorrect. When Lee prepares a manuscript on his work, Dr. Johnson suggests that Dr. Smith be included as a co-author. Lee counters that Dr. Smith was not involved in the research, but Dr. Johnson argues that Dr. Smith supported Lee's salary and provided some of the reagents. When Dr. Smith reviews the manuscript, he demands to be made senior author and to have the data removed that do not support his model. Lee and Dr. Johnson are not willing to make these changes. When Dr. Smith does not agree, Lee and Dr. Johnson remove Dr. Smith's name from the paper and acknowledge his support through the training grant and provision of reagents.

Key Words Are Important to Keep Your Work from Being Lost

Key words are often provided at the last minute when submitting a manuscript. They are very important since the index of the journal is based on key words and they may influence some search engines. Spend time making a list that may be too long at first, and then select the best from the list to the maximum number allowed by the journal. Newer search engines examine the full content of the article's text, but do not ignore the key words.

Your Abstract Is Second Only to Your Title as an Opportunity to Attract Readers

Each journal will have specific requirements regarding the length, content, and format (e.g., structured or unstructured) of abstracts. Abstracts should be written for a general audience and should briefly describe the background and purpose of the study, methods, results, discussion, conclusions, and speculation. You should use the guidelines we discuss earlier for preparing abstracts for scientific meetings. You should avoid abbreviations and jargon.

Search engines such as PubMed provide the title, followed by the abstract. Therefore, it is important to have an abstract that will attract the searcher to your full text.

The Introduction Sets the Stage in Three Subparts

The introduction provides the background for your research. It specifies the questions and goals of the research and shows how these developed from previous research. You should write your introduction so that the reader considers your research the next necessary step in the field.

The introduction typically consists of three subsections, although these will not usually have subheadings. The first, usually a single paragraph, describes the contextual significance of the work and defines any terms that may be critical to the study. The second subsection is usually one or more paragraphs (rarely more than three) in length and provides the historical background and previous literature directly relevant to the primary topic or purpose of this work. For this reason, this second subpart of the introduction is more heavily referenced than the first or third, and it is important not to bring in material that more correctly belongs in the discussion. The third subsection describes the purpose of the study and may include a very brief description of the results and, therefore, is a partial abstract of the work, containing the core message of the paper: its hypothesis (purpose) and a brief data summary.

Your Methods Section Should Represent a Logical Experimental Approach

The methods section should be written in enough detail that anyone can repeat your study. It is best to break this section up with specific subheadings organized according to the design or logic of the study.

If the study involves human subjects, consider the following. Approval by the institutional review board is required for human subjects. You should include a description of the relevant attributes of the human participants in your research, including age, gender, and ethnocultural group. If the work requires one or more case descriptions, these should be provided in a separate section, usually immediately before or after Methods or as the first subheading in this section. You should follow the journal's instructions. Patients should not be identified by name. Written consent is required for photographs, and the individual should not be identifiable. Any information that the individual or a family member might find inappropriate (e.g., misattribution of paternity) should not be included. It is acceptable to fictionalize a pedigree as long as this is noted, and no reason need be included. Pedigrees should be altered to prevent identification of family members while maintaining the genetic information, and again, this should be noted.

If animals are used, consider the following. Animals should be described by species, strain, age, sex, and other important characteristics and you should have approval from the animal research committee.

Refer to other papers for published methods. If you modified published methods, cite the original and describe the changes you made. You should specify the manufacturer of unusual chemicals, reagents, and equipment.

Modify Pedigrees to Present the Inheritance Pattern but Not the Status of Individual Family Members

Lynn was preparing a manuscript on a rare disease. As part of the patient information, he planned to include the pedigree of a family that was very important to his research. When he reviewed the Instructions for Authors, he found that the editorial policy of the journal was to restructure family pedigrees so that no one would be able to identify the family member who had the disorder or members who were carriers for the disorder. The rationale was that there may be family members who had not participated in the research or did not want to know their genetic status regarding disease or carrier states. Participants in the study may not want others to know their status. For a rare disorder, a family pedigree could identify the family and the individuals within it. To conform to the editorial requirement and to maintain the confidentiality of family members, Lynn constructed an idealized pedigree that conveyed the typical inheritance pattern but would not identify the family or any individual member. Lynn noted that this pedigree was idealized to maintain confidentiality.

The Results Section Should Build Rationally toward Your Logical Conclusion

The Results section describes data summarized in tables and figures. The results may be organized by subheadings according to the study design. You should not present the same data in the text as table(s) or figure(s) but should describe and summarize the information in the table(s) and figure(s). The statistical tests used should be specified (e.g., standard deviation or standard error of the mean). You should be rigorous in your statistical analysis. For example, there is no such thing as "almost significant" or a "trend" toward significance; you set the level of significance required a priori, and your data either meet the level of significance or they do not.

You should use color figures if they convey information that would be lost in black-and-white figures. Color figures add significantly to the cost of publication. Authors are often asked to pay a certain amount for each page of text and for each color figure.

Results That Are Almost Significant Are Really Not Significant

Sam prepared a manuscript describing her research results. Before performing the research, she had set the significance level at 0.05. When

Sam conducted the research and performed the statistical analysis, she found that one of the twenty tests was statistically significant at the 0.05 level. Three of the twenty tests achieved significance levels between 0.05 and 0.10. Sam presented the results and went on to discuss them as if all four of these tests confirmed the hypotheses they were designed to test. Sam was surprised when her manuscript was rejected. Reviewer 1 pointed out that Sam should discuss only the single result that reached the level of significance that she had set at 0.05. Reviewer 2 described Sam's results as inconsequential. If one test out of twenty was statistically significant, this is only the number expected by chance. The editor of the journal agreed with both reviewers and rejected Sam's manuscript.

Expound Rationally and Succinctly in Your Discussion

The Discussion section provides an interpretation of findings and their importance within the context of the literature. Typically, the Discussion is organized into three subsections. The first, usually one to several paragraphs in length, summarizes the results of the study, logically developing the argument for how the data address the hypotheses and fulfill the purpose of the study. The second subsection describes how your work fits into that of others and contributes to the knowledge in the field. This section may be as long as necessary. You should describe the limitations of previous research and how your current findings improve or clarify the issues. Specify how your research confirms previous findings or provide possible explanations for results that differ from previous work. The major limitations of your study should be specified. The third subsection is usually only one paragraph long and provides a capsule summary of the study conclusions, and any speculations or impact on clinical practice, as well as any future directions for your work that you may wish to discuss. Be cautious not to overstate your conclusions or to speculate excessively.

All Data Need to Be Included and Placed
in the Context of Previous Models

Three faculty members, Dr. Davis, Dr. Jones, and Dr. Smith, were collaborating on research that was an extension of a model developed by Dr. Jones. The data contributed by Dr. Davis and Dr. Smith clearly demonstrated that there needed to be a change in Dr. Jones's model. Dr. Jones insisted that Dr. Davis and Dr. Smith repeat their experiments. Dr. Davis and Dr. Smith replicated their previous results. Dr. Jones refused to accept this challenge to her model and threatened to block any attempt by Dr. Davis and Dr. Smith to publish their work. Dr. Davis and Dr. Smith consulted the department chair. The department chair made it clear to Dr. Jones that the manuscript would include Dr. Davis's and Dr. Smith's data and would discuss changes in Dr. Jones's model. The department chair said that Dr. Jones needed to decide if she wanted to be included as an author on this manuscript.

The Reference Section Should Be
Checked Carefully for Accuracy

You should follow the journal format for the references and be sure of the accuracy of the reference citations. To check the accuracy of references, go back to the original articles yourself. It is amazing how often an error in a reference is repeated over and over, showing that a long series of authors were relying on the citation and not reading the original article. You should also visit the library or use interlibrary loans for sources that are not available online. You don't want to rediscover something that was well known twenty years ago. Also, clarify the journal's policy for citations to in-press, submitted, in-preparation, and personal communication references. These policies vary among different publications. Some journals require files containing manuscripts that are submitted or in press (if they are not available online). Some journals require a letter from the person cited as personal communication.

Use Primary Resources

Lee was in a hurry to prepare a manuscript. Rather than perform a thorough review of the literature, Lee read a recent review and used the citations from the review. One of the reviewers of Lee's manuscript was an author of one of the articles Lee cited. The reviewer noted that Lee described the reviewer's results incorrectly and listed the reference to the reviewer's study with the wrong volume and page number. Since Lee had also cited the review article, the reviewer looked at the review article and noted that Lee had used the reference citation from the review. The reviewer correctly assumed that Lee had not read the original article. The reviewer summarized these findings in both the confidential comments to the editor and the comments to the author. The reviewer recommended rejection of Lee's manuscript, and the editor agreed. In an attempt to save time, Lee had actually wasted time by not undertaking a thorough review of the primary literature. Lee's reputation was also damaged.

Good Figures Attract Attention to Your Work

Tables and figures should contain enough information to stand on their own. You can use figures to sell your manuscript, especially if you invest in color. Some journals have the policy of accepting one or more color figures per article free of charge. A number of journals have a color figure on the cover of each issue. By submitting a color figure with your manuscript, you are in the running for the cover. If your figure is selected, you should request a copy of the issue, even if you do not subscribe to that journal. You may want to frame the cover to hang in your office or lab. You should also list the reference in your curriculum vitae as the full reference and add "with cover." A cover figure is an excellent vehicle for marketing your research.

Many journals limit the number of tables and figures for an article. They typically allow supplemental tables and figures to be available online. You should consult the journal's policies in the Instructions for Authors.

HOW TO WRITE REVIEW ARTICLES AND CHAPTERS

Strategically Consider Opportunities for Review Articles

Throughout the course of your career, you may receive numerous requests to write review articles and chapters. Criteria for promotion and tenure, as well as reviews of grant proposals, focus on first-authored or last-authored, peer-reviewed publications. Review articles and chapters are given relatively little consideration during evaluations for those in tenure-track faculty series, though they may be more important in clinical or master teacher series. In the tenure track, these review articles and chapters will be important only as they are evidence of peer recognition. Therefore, it is important to publish in the most prestigious journals and books in your area. You do not want to spend time writing review articles and chapters, because they keep you from having the time to write peer-reviewed manuscripts.

One legitimate reason to write a mini-review or review is to get new ideas and/or models into the peer-reviewed literature. This will give these concepts more credibility when you cite them in your grant applications.

Expect Authorship for Work You Author, and Ask at the Outset

Dr. Smith called graduate student Lynn into the office. Dr. Smith was excited about a request from the editor of a prestigious series to write a chapter. Lynn congratulated Dr. Smith on this opportunity. Dr. Smith then suggested that Lynn write the chapter. Lynn was thrilled about this opportunity. However, when Lynn asked about the order of authorship, Dr. Smith said there would be only one author, Dr. Smith. Stunned, Lynn asked if Dr. Smith meant that Lynn would do all the writing and only Dr. Smith would get credit for Lynn's work. When Dr. Smith replied that Lynn was correct, Lynn politely but firmly declined to have anything to do with the chapter. Dr. Smith was unwilling to grant Lynn authorship and would either have to write the chapter as sole author or find someone else to be a "silent partner." Dr. Smith would not have been diminished by Lynn's co-authorship and it would have been a real boost for Lynn's career.

Write about What You Know

Restrict the topic of review articles or chapters to your research area or clinical subspecialty interest. You should focus on an area that you want recognized as yours. You should already have sufficient knowledge and background in this area to provide a strong foundation for your review of the topic. It should be an area in which you have a reputation or in which you want to develop one.

Authorship on Chapters Can Be Changed

Dr. Jones, the department chair, was editor of one of the most prestigious books in the field. Dr. Smith, a junior faculty member, was flattered when Dr. Jones asked her to become a co-author of the chapter in her field. Dr. Jones commented that she would have a lot of work to do to update the chapter. She was honored to do so and to be a co-author with the more senior faculty in her field in the department. The more senior faculty agreed with her changes. Dr. Smith received an excellent offer to move to a different institution. She was dismayed when the next edition of the textbook came out with no changes to her chapter except for the removal of her name as a co-author. She contacted Dr. Jones, who replied that one had to be a member of the department to be a co-author.

Turn Work You Have Already Done into Publications

Any review articles or chapters you write should be based on your multiple research articles or clinical reports, your recent grant application, or your recent one-hour talk. It is important to take your existing, but unpublished, work and turn it into a product that will reach a broader audience through publication. The timeliness of a review article is determined by a controversy in the area, your ability to uniquely organize the available information, or the need to direct future research efforts.

Peer-Reviewed Research Articles Should Take Priority

Dr. Jones was flattered that Dr. Smith, a senior member of Dr. Jones's department, asked her to write a chapter for a book he was editing. The topic was outside Dr. Jones's research area, but she decided the extra research needed to write the chapter was worthwhile. She delayed the preparation of several research manuscripts in order to do the literature search and write the chapter. She submitted the chapter on schedule. Six months later, she realized there had been no communication from Dr. Smith regarding the book. When she asked Dr. Smith, he said the publisher had gone bankrupt and the book would not be published. When she went up for promotion, she was criticized for having too few research publications. Not only was the chapter not on her curriculum vitae, but also the work on the chapter had delayed preparation of critical research publications. You can imagine her dismay when two years later, Dr. Smith told her that there was good news about the book. A new publisher had acquired the rights to the book and wanted to publish it. Dr. Smith said that Dr. Jones would have to update her chapter. She debated whether she should withdraw the chapter or put in a lot of work to update it. She chose a compromise. She added a few new references with a minimal amount of rewriting. She also vowed not to agree to write any chapters unless the publisher was reputable, the editor had an extensive track record of successful books, and the chapter was in her area of expertise.

Know Early in the Process Whether a Journal Is Interested in Your Review

There are a number of criteria you should use to select an appropriate journal for your review article. If the editor of a journal requests a review from you, you need to decide if the topic fits your area of focus, if the journal is one in which you want to publish, and if your article will be accessible via the online search engines for your discipline. Journal editors solicit reviews as a way to increase their impact factor, since they are cited more frequently than original research articles. This is the rationale behind the founding of journals that publish only reviews.

If a journal has published multiple research papers or clinical reports on this topic, contact the editor to determine whether there is an interest in your proposed review article. You could also contact the editor of a journal that targets an audience you would like to reach with a review article. Factors to weigh in your selection of a journal include the reputation of the journal and the speed of online publication.

KNOW THE REVIEW MECHANISM FOR THE JOURNAL YOU SELECT

There are various types of review mechanisms. Some editors are very hands-on and typically assign reviewers and make a decision based on the reviews. When you submit your manuscript to a journal with a different review approach—for example, multiple communicating editors—you select the most appropriate editor, who then may choose the reviewers and make the decision regarding your manuscript. In some cases, the final decision will be made by the editor-in-chief or by a weekly meeting of the editorial board.

Have a Timeline for the Review Process in Mind and Keep Track of It

Lee just made an important research breakthrough in an area in which a number of groups were in intense competition. Lee submitted a manuscript describing her discovery to a prestigious journal and requested an expedited review. The editor agreed to the expedited review. While reading the manuscript to decide on potential reviewers, he realized that Lee had just accomplished what his friend had been working on for several years. The editor contacted his friend and told him about Lee's manuscript. The editor slowed the review of Lee's manuscript to allow his friend to complete his work and submit a manuscript to another journal. Once his friend's manuscript had been submitted, the editor accepted Lee's paper for publication. While she was concerned that her review had not been expedited, Lee was relieved to have this breakthrough published. When the editor's friend's manuscript was published a month later, Lee had no way of knowing how the editor had intervened. Perhaps if Lee had pressed the editor for an expedited review when the schedule specified for such reviews was not followed, he may have been forced to process her manuscript more rapidly. Lee should have withdrawn her manuscript and submitted it to another journal when the editor did not provide the agreed-upon expedited review.

Retain a Sense of Control over Your Work during the Review Process

Recognize that you are not completely powerless during the review process. If the evaluation and review are taking longer than you would like, and longer than is usual for that journal, you may withdraw your manuscript. You must do this before submitting it to another journal.

Remember that there are certain times of the year when it is very difficult to get reviewers for any manuscript. The month of August and the weeks between Thanksgiving and New Year's Day are particularly difficult. In editing our journal, we have noted that there is often an inverse relationship between the willingness to review manuscripts for the journal and the demand that one's manuscript be reviewed quickly. Editors remember who publishes in their journal and who reviews for their journal. They also remember the difficult authors. Being a willing, timely, and knowledgeable reviewer is part of establishing your professional reputation.

If You Write, You Must Be Willing to Review

As an assistant professor, Kim is aware that peer-reviewed publications are important to her promotion and tenure. She submits manuscripts at the rate of one per month. Her publication rate has impressed the senior faculty in the department. Kim is often asking editors to expedite the review of her manuscripts. However, she never agrees to review a manuscript. Finally, one editor, who had published several of Kim's manuscripts, was frustrated by her fifth refusal to review a manuscript for the editor's journal. The editor spoke to Kim and reminded her that reviewers were an important part of the journal infrastructure. If everyone behaved like Kim, there would be no peer review of manuscripts.

Know the Possible Review Outcomes and Their Implications

Always read the letter from the editor carefully in addition to the reviews. The editor may have additional comments and may also indicate that you should emphasize certain parts of the reviews in your revised manuscript—for example, to focus on reviewer 1's comments more than reviewer 2's. The editor decides what happens to your revised manuscript, so, if at all possible, you should try to make the changes suggested by the editor. If the editor does not give you guidance, you should try to comply with the majority of the reviewers' comments. If you ignore all of the reviewers' recommendations, you may find that the editor rejects your "revised" manuscript.

The Editor's Comments Give You an Assessment
of the Reviewers' Comments

Sam received two reviews of his manuscript. One was relatively positive with a few minor changes recommended. The other reviewer was very negative, requesting major changes, additional experiments, and a significant reduction in the length of the manuscript. In his comments to Sam, the editor suggested that he revise the manuscript without performing additional experiments or shortening the manuscript. Sam made the changes that were reasonable without doing additional work or significantly rewriting the manuscript. When describing the response to the editor's and reviewers' comments, Sam explained why the additional experiments requested by the second reviewer were unnecessary. The editor accepted Sam's revised manuscript.

You may be fortunate enough to have your manuscript accepted without any changes on its initial submission. However, this is an extremely unusual outcome. Usually some revision is required. If the editor asks you to revise your manuscript, you should make the changes you are comfortable with and present reasons why you are unable to make the other changes. You should meet the deadline the editor sets for your revised manuscript. You should respond to each suggestion by the editor and the reviewers. The editor doesn't have the time to compare your original and revised manuscripts to determine what changes you made. You should list each comment of the editor and the reviewers and describe your response.

Your manuscript may be rejected, but the letter from the editor may suggest that you follow the reviewers' suggestions and submit your revised manuscript as a new paper. This increases the rejection rate and the submission rate and decreases the turnaround time for the journal, all of which are important measures used by the publisher and the editor to gauge a journal's success.

Sometimes when the editor rejects the manuscript, this is the final decision. If this is the case, it does little good to argue with the editor. You would be better off to make the edits in the manuscript suggested by the reviewers and submit your revised manuscript to another journal.

Do Not Let Your Pride Interfere with Publication

Sam made an important discovery and submitted a manuscript to the top journal in the field. Sam was disappointed when the editor requested making several minor changes and shortening the manuscript to conform to the requirements for a brief communication. Sam submitted the full-length manuscript to another journal. While the second journal did accept the manuscript as a full-length article, it was published six months later than it would have been if Sam had agreed to shorten his manuscript for the original journal. In retrospect, Sam wondered if he had made the right decision, since the second journal had less prestige than the first and a readership that was not as broad.

Distinguish between Perseverance and Perseveration

If your manuscript has been rejected by three or more journals, you need to decide if you are persevering or perseverating. You should ask your mentor to provide an honest opinion of how you should proceed.

Pay Careful Attention to Reviewers' Comments

Chris was a new faculty member who was feeling a great deal of pressure to publish. Chris was wise enough to ask several senior faculty members to review a manuscript before submission. All of the faculty members agreed that Chris needed much more data to support the conclusions. Chris dismissed their suggestions and submitted the manuscript. The first journal rejected it, and the two reviewers made points similar to the senior faculty members' suggestions. The second journal returned the manuscript to Chris without review, saying that the manuscript was inappropriate for that journal. The third journal had one positive review and one negative review. The editor of the third journal rejected the manuscript. Chris called the editor of the third journal and requested a third reviewer, since there had been one positive review. The editor agreed to contact a third reviewer. The third reviewer had previously reviewed Chris's manuscript for the first journal. In the comments to the editor, this reviewer was adamant that the manuscript be rejected and that the author be told to make the changes the reviewer had recommended in the initial review for the first journal. By refusing to listen to the advice of senior faculty members, reviewers, and editors, Chris was on the way to establishing a negative reputation. Senior professors would be unwilling to review future manuscripts, because they would feel that this was a waste of their time. When reviewing Chris's future manuscripts, abstracts for scientific meetings, and grant proposals, reviewers would remember Chris's lack of substance. Editors would remember Chris's lack of responsiveness when deciding whether Chris's future manuscripts were appropriate for their journal, and when considering potential reviewers for manuscripts in this field, they would not consider Chris.

RESPONDING TO THE REVIEWERS' COMMENTS

Make the Easy Changes and Negotiate the Tough Ones

Read the reviewers' comments carefully and make all the changes you can to improve the manuscript. These include following the Instructions for Authors; following spelling, grammar, and nomenclature conventions; making stylistic changes; including additional information when available; including additional data when feasible; and restructuring the discussion to truly reflect the data.

Authorship Can Be Changed during the Review Process

Lee is a student doing research. A senior faculty member in the group asks to see Lee's data. Another figure is needed to respond to reviewers' comments on a manuscript. Lee asks to see the manuscript and is dismayed to find her data as the only data in the paper and her name missing from the list of authors. Lee goes to the department chair, who immediately recognizes the error and demands that Lee be made a co-author.

There are some changes that you may find difficult to make. Are you willing to provide additional data when months of work are required, or should these data be the basis for your next manuscript? Editors are easily frustrated by authors who submit a paper containing "the least publishable unit"—that is, the minimal data they think will support a manuscript. Often reviewers are asked to give a publication priority rating for the manuscript. Editors are not going to publish low-priority papers with too little new information.

Are you willing to restructure your paper to provide support for a different point of view? Some editors may ask you to change the fundamental message of your manuscript. You and your mentor(s) should determine if such a change is warranted, or if the editor represents an opposing perspective in a controversial area and you should submit your manuscript elsewhere.

Know When to Move On to the Next Journal on Your List of Three

Sandy submitted a manuscript to a very prestigious and rigorous journal. Reviewer 1 did not agree with Sandy's interpretation of the data and wanted Sandy to rewrite the discussion to reflect reviewer 1's point of view. Reviewer 2 felt that Sandy did not have sufficient data and needed to perform a series of experiments that would take about six months to complete. The editor stated that Sandy's revised manuscript would require re-review. In spite of the journal's reputation, Sandy decided not to take the chance that further research would provide the results anticipated by reviewer 2. Sandy was also concerned that reviewer 1 had misinterpreted the results. Sandy was unwilling to change the discussion to reflect reviewer 1's opinion. Sandy was also concerned that if the revised manuscript was re-reviewed by the original journal, it still might not be accepted. Sandy revised the manuscript in accord with the Instructions for Authors for her second-choice journal and made the changes she could that were recommended by the reviewers for the first journal. Her paper was published within six months.

Your Communication Accompanying Your Revised Manuscript Should Make the Editor's Job Easy

Copy and paste the editor's and reviewers' comments into your letter responding to the editor and reviewers. Under each point, provide your response in clear language. You should not be argumentative or defensive. If you are unwilling to follow a recommendation, give your reasons. Try to accept more suggestions than you reject. Thank the editor and the reviewers for their comments and tell them that their recommendations improved the clarity of the manuscript.

HOW REVIEWERS ARE SELECTED

Editors select reviewers from leaders in the field. They are likely to ask individuals who have provided good reviews in the past, who publish in the journal, and who are members of the editorial board. They may contact scientists they have met or heard present at meetings. They may do a keyword search to determine who publishes in the area, or they may use the authors of references cited in the manuscript for an author search. Editors may also use the reviewers suggested by the author after they determine if they really have expertise in this area and if there is no conflict of interest. Conflict of interest would occur if the reviewer and an author were collaborators and had published together in the last five years, if the reviewer and an author are from the same institution, or if the reviewer uses the review of the manuscript to obtain an unfair research advantage.

Some potential reviewers say they can't review a manuscript because they publish in the area and it would be a conflict of interest. Who would make a better reviewer than someone who does research in that area? A conflict of interest would arise only if the reviewer was late with the review or was inappropriately harsh in their criticism of a rival's manuscript. To avoid any conflict of interest you should agree immediately and submit your fair review of the manuscript that day.

Obtain Credit for Your Efforts as a Reviewer

Sandy is a postdoctoral fellow in Dr. Jones's lab. Dr. Jones is an international expert in the field and is often asked to review manuscripts. Even though she is very busy, Dr. Jones always agrees to review manuscripts. She doesn't review them herself, but instead has the students and fellows in her group review them. Dr. Jones never tells the editor that a student or fellow did the review; she claims the credit for herself. Sandy is often asked by Dr. Jones to review manuscripts. Sandy enjoys the review process and Dr. Jones praises Sandy's reviews. Sandy is curious about each manuscript he reviews and always reads the paper if it is published. The most recent paper Sandy reviewed is accepted for publication, and Dr. Jones is asked to prepare a commentary to appear with the article. Suddenly, it dawns on Sandy that Dr. Jones used him to write the review that became the basis for the commentary. Dr. Jones took credit not only for Sandy's work in doing the review but also for the published commentary based on his review. The next time Dr. Jones asked Sandy to review a manuscript,

Sandy stated that he would do the review only if Dr. Jones told the editor that Sandy was the reviewer and if Sandy could communicate the review directly to the editor. Faced with Sandy's ultimatum, Dr. Jones agreed.

HOW TO REVIEW A MANUSCRIPT

Accept Your Responsibilities in the Review Process

When you are asked to review a manuscript, accept if you can meet the deadline, if the topic is within your area of expertise, and if there is no conflict of interest. You should reserve your comments for substantive issues. If the writing is so poor that it interferes with comprehension, you should include this information in your review. You may mention examples of poor writing in your confidential comments to the editor and suggest that, if accepted, the manuscript will require extensive copyediting. If the writing raises only minor issues, the copyeditor's routine process will deal with spelling, grammar, and style if the manuscript is accepted for publication.

You should return your review in a timely fashion. Remember how frustrating it can be when your manuscripts are delayed due to slow or unwilling reviewers. Some editors may ask you to re-review manuscripts that have been revised. We do not burden our reviewers with this task and feel that the editorial office should be able to determine if the revised manuscript has met the standards of the reviewers and the editor.

Information in the manuscript is confidential. You may not cite it until the manuscript is accepted and available online. Many editors will let you know the decision regarding the manuscript at the same time the author is told. If you review regularly for a journal that does not notify reviewers of the outcome, you might suggest that the editor extend this courtesy.

Editors remember good reviewers when they submit manuscripts of their own or when it is time to add someone to the editorial board. Editors are also likely to serve as reviewers of grant proposals or abstracts for professional meetings. Serving as a reviewer or editor allows you to help shape your field—publishing good work and keeping bad science out of the literature.

HOW EDITORS ARE SELECTED

Editors Are Selected by a Variety of Mechanisms

Some journals elect their editors. They are nominated by their colleagues and elected by the membership of the organization that supports the journal. Other editors are selected by the leadership of the professional organization represented by the journal, or from the editorial board of the journal on the basis of their abilities as reviewers and their commitment to the journal. Some journals advertise for the editorial position before the editorship is vacated. If you are interested in such a position, ask your senior colleagues if they think you would have the requisite credibility and whether they would be willing to write a letter of nomination on your behalf.

Becoming an Editor Requires Preparation

Lee is an ambitious assistant professor. One of his goals is to become the editor of a major journal. Lee decides the first step toward achieving this goal should be membership on the editorial board of a major journal. He e-mails the editors-in-chief of several major journals, requesting membership on the editorial board. In support of his request, he attaches his curriculum vitae. While Lee has a number of publications, he rarely agrees to review a manuscript. Some of the journals Lee contacts are ones for which he refused to serve as a reviewer, consistently stating that he was too busy. Most of the editors simply delete Lee's e-mail. One takes the time to write back and explain that members of the editorial board are selected based on extensive, skilled reviewing for the journal and an international

*reputation in the field. In addition, editorial boards need to have geographic
diversity and represent each of the topic areas covered by the journal.*

WRITING BOOKS

In some fields, single-author published books are the currency for
tenure and promotion. Except for phenomenal best sellers or widely
adopted textbooks for introductory courses with large enrollments,
don't expect to make a lot of money publishing books. You should
write a book to advance your field, to promote your ideas, and/or to
educate.

For other fields, edited volumes take the place of, or supplement,
peer-reviewed journals. The editor or editors are responsible for
selecting the topics and authors, editing the text, and ensuring
timely delivery of the chapters to the publisher. These are even less
likely to make a lot of money, unless they are widely adopted textbooks. These are often the result of a conference in which a commercial sponsor agrees to support publication of the edited volume.

Book contracts typically provide a relatively small percentage of
the proceeds to the author(s) as royalties, less any advance made to
the author(s). You may want to limit the number of figures and tables
in your book because you may be responsible for paying for these
out of an advance on your royalties. You need to read your contract
carefully and be sure you understand each aspect. One point that may
be confusing is the lack of royalties for books sold at a discount. A
vendor or private party can negotiate a reduced price for a large number of copies. This means you sell more books, but you do not receive
a royalty for these copies. Another issue is translation of your book.
Some publishers do not provide royalties for translations, so they are
paying for the translation using money that would otherwise come
to you as royalties. As more publishing becomes electronic, publishers
may market electronic versions of part or all of your book. Be sure
you understand the implications of this for your royalties. One pub-

lisher recently offered to print individual chapters in its journal so they would be accessible to online searches. However, the copyright and royalty issues were unclear.

Should you use an agent? It would be ideal to have a knowledgeable individual negotiating a contract on your behalf with the publisher. However, agents are interested only in potential best sellers, especially by authors who have already had a best seller. Most academic authors are left on their own. Certainly any advice from your mentors about this issue would be helpful.

If you are writing a book for the first time, you will need to prepare a working title, a detailed table of contents, and one or more complete chapters. You should contact appropriate editors at publishing houses that have a commitment to your area. You will be more successful if you already know the editor or if a colleague can provide an introduction to the editor. The editors will help you determine if there is a need for such a book and if the publishing house has an interest in your proposal. In your discussions with the editor, you should be prepared to describe the market for your proposed book and whether you are writing for the general public (trade) or students (textbook), or both.

If you prepare a complete manuscript before you begin this process, you may find that you have put in a lot of work that does not have a market or that needs to be totally rewritten to be of interest to a publisher.

Your Book May Evolve over Time

Dr. Johnson had an idea for a textbook for a course that she had taught for three years. A colleague introduced Dr. Johnson to the editor of his textbook. The editor was impressed with Dr. Johnson's concept, working title, detailed table of contents and complete chapter. With the editor's encouragement, Dr. Johnson completed the book and forwarded it to the editor. The editor responded that her book was a cross between a trade book and a textbook. Since the publisher dealt only with textbooks, the editor's company would not be the best at marketing Dr. Johnson's book. The editor was kind enough to introduce Dr. Johnson to editors at three different publishing houses that regularly produced trade/textbooks. All three editors expressed interest in Dr. Johnson's book. The first editor asked Dr. Johnson to completely rewrite her book to conform to a position in the field that was diametrically opposed to hers. She declined to pursue this possibility. The second editor was very enthusiastic about Dr. Johnson's manuscript and wanted to move forward to a contract. When Dr. Johnson contacted the third editor to say that she was expecting a contract from the second editor, the third editor could only apologize for a colleague who had dropped the ball. Dr. Johnson moved forward with the second editor, who proved to be very helpful and supportive during the publication and marketing of her book.

CHAPTER 6

Gender Issues

While those of us who are women have come a long way, we may occasionally confront situations in which it would appear that our gender is holding us back. When these arise, it would be wise to consult one or more mentors to determine if this is truly the case. Should your mentors confirm your suspicions, you and your mentors should determine whether to confront the individual or to address these issues with your university ombudsperson. If your mentors disagree with your perception, you need to ask yourself if you are viewing all aspects of your career through gender-colored glasses. Do you interpret every bump in the road as being due to prejudice against women? If so, you need to step back and take responsibility for your own behavior and not consistently misinterpret the responses of others.

THE IMPORTANCE OF A POSITIVE ATTITUDE

Women truly need to love themselves for who they are. They need to celebrate their good qualities, work on their weaknesses, and accept responsibility for their life choices. Women need to reach out to other women and support them without making a value judgment.

Women need to be confident enough in their own life choices that they can respect others' different life choices. Women need to accept—even cherish—the mentorship of men who have a successful track record with mentoring women and who are willing to treat women as their intellectual equals.

Unfortunately, women all get painted with the same brush. If a guy screws up, he's just a jerk. If a woman screws up, the response is often "What do you expect? She's a woman." Women need to consider their value to the department and the organization. They need to be sure they are meeting not only their own goals but those of their mentors and department chairs as well. If they are meeting all of these goals, they may need to remind their department chairs that they are doing so.

BENEFITS OF SUPPORT PROGRAMS FOR WOMEN

One thing that some women lack, just like some men, is an understanding of teamwork. One goal of women's support groups is to foster a sense of teamwork among the members. It may be necessary from time to time to sacrifice individual goals for the team. These personal sacrifices on the part of one team member will ultimately benefit all members. Teamwork becomes ever more important as one progresses through one's career.

Women's support groups can also provide training in other unwritten skills necessary for academic achievement and ultimate career success. These skills include those involved in selecting mentors and collaborators, presenting talks, preparing manuscripts, writing grants, developing research collaborations, teaching, pursuing leadership opportunities, and balancing personal and professional lives.

These groups should also provide insight into institutional support for pregnancy, maternity, and motherhood. They should lobby the administration for increased support for parental leave, day care, before- and after-school care, and vacation child care.

Let's face it, there is no good time to get pregnant during your career. Your insurance should cover your pregnancy. You should be allowed to take medical leave for complications during pregnancy. You and your partner should be allowed to take parental leave after your child is born. There are grants to support training for those who have taken time for their family and want to re-enter academics. Faculty at many institutions can add one year onto their tenure clock for each child they have while on the faculty. If, while you are on faculty, your institution reviews you for promotion to associate professor and tenure during your sixth or seventh year and you have two children, your tenure decision can be deferred until your eighth or ninth year on faculty.

PROBLEMS WITH SUPPORT PROGRAMS FOR WOMEN

Should a support program for women evolve into simply an opportunity to complain, this not only is a waste of time but is also counterproductive. The attitudes encouraged by such groups reinforce negative stereotypes about women professionals. Imagine our dismay when we were doing a workshop with a women's group that became very unproductive when the members began berating us as if we had caused their problems instead of discussing possible solutions. Organizations that support the attitude that anyone outside the group is the enemy will not lead to any useful outcomes.

INABILITY TO MAKE GENDER-NEUTRAL CHOICES

Some women choose an exclusively women's college to develop their leadership and academic skills in a supportive environment. They derive a great deal of confidence from such an opportunity, which they bring to their further training and/or profession.

Consistently choosing female professors, mentors, and professionals because they are female may not always be the best choice. No single characteristic, even gender, should be the only one taken into consideration.

TRANSLATING YOUR VALUES INTO ACTION

Some women want to have it all—a family and a tenure-track academic position. With only 20 percent of PhDs achieving the goal of a tenure-track position, it would appear to be difficult with or without a family, regardless of gender. Certainly a significant number of women have been able to accomplish these goals. Were they smarter? Did they work harder? Did they have better mentors? Were they at the right place at the right time? Did they make better professional choices? Were they luckier? Perhaps they were more likely to realize that life is a marathon, not a hundred-yard dash, taking more time during training to be sure they had established the foundation for their faculty career with experience in presenting research, preparing manuscripts, and writing grants. This strategy is effective regardless of gender or parenthood.

Extending Fellowship Training

Christine is an MD who wanted a family and a tenure-track position at an academic medical center. With the demand of medical school and residency, she decided to postpone childbearing until her fellowship years. She extended her fellowship training for three additional years so that she could have two children and develop the skills necessary to launch her academic career. With support from her partner and her mentors, she has risen to the top of her specialty, and her research receives international recognition. While her salary during the three additional years of fellowship was not as much as what she could have earned on faculty, she more than made up the difference through her rapid rise through the ranks.

"LESS" CAN BE MORE

Most major universities are more interested in their bottom line than in their academic mission. They want tenure-track faculty who can bring in grants, develop patents and companies, run profitable services, and/or court donors. If you are a superb teacher or active in the community, that is a nice bonus, but not a requirement. To have enough time to bring in money, faculty may minimize their teaching load. Courses are often taught by part-time adjunct faculty, post-doctoral fellows, or graduate students. Mentoring may be discussed in passing, but there is little or no recognition of successful mentoring beyond applications for training grants.

As a woman, your values may tend more toward teaching, mentoring, and public service. You may find that there are certain academic institutions, government agencies, private companies, and private foundations that share your values. You should seek a position with them. In the end, you will be more successful in a position that is a good match with your values. Be true to yourself.

Commitment to the Underserved

Sarah knew from the first day of kindergarten that she wanted to be a teacher. In college she fell in love with the discipline of anthropology and decided to become a professor in that field. During graduate school she enjoyed her research, but her opportunities to be a teaching assistant were the most rewarding part of her training. As the first in her family to attend college, she had a commitment to the underserved. When she landed a faculty position at the urban campus of a state university, she found her niche. She didn't mind the heavy course load. It gave her the opportunity to reach out to more students.

EQUAL OPPORTUNITY

Educational opportunities, training slots, fellowships, and jobs should be awarded to the most qualified individuals, regardless of gender. Applications for educational and training opportunities are available on institutions' websites. Fellowships and jobs are advertised nationally on the Web and through more traditional print sources. The purpose of this openness is to provide equal opportunity. Institutions do track faculty hiring and promotion practices by gender. The goal is to have parity in hiring, promotion, and salary.

Concern Regarding Pay Raises

Susie, Kathie, and Keith were hired at the same time. Susie and Kathie were in the English Department and Keith was in the History Department. At the end of their first year of teaching, Susie and Kathie felt they had each worked very hard and had exceeded the goals set for them by their department chair. When their contract for the next year showed that each would receive a raise, they were concerned that the raise was too small. They wondered if Keith was getting a more substantial raise. When they asked Keith, they found out that his raise was half of theirs. After this, their relationship with Keith was never the same, because he remained envious of their raises.

LACK OF MOBILITY OR INOPPORTUNE MOBILITY

Couples in academics are referred to as "the two-body problem." If one member of a couple is highly sought after by an institution, the institution typically circulates the CV of the other member to relevant departments to determine if there is any interest. If there is interest and an opening, things might work out well for both partners. If there is no interest and/or opening, the couple needs to consider their options and make a decision based on their current situation and other opportunities. Larger metropolitan areas with multiple institutions and opportunities for both partners are more attractive to couples.

There have been occasions when departments have taken advantage of women in relationships with partners who were tied to the area. These women may not be paid or promoted fairly because the department expects that the women won't be leaving under any circumstances. This should be picked up by the annual surveys of salaries and promotions and dealt with by the dean. If it isn't, the female faculty member should consult with her mentors and develop a plan to meet with the department chair, the dean, or the ombudsperson.

Sometimes one member of a couple has settled into a good situation and it is time for the other member to move on to a different setting. Both members of the couple should consider their options and come to a joint decision. Like the relationship itself, this is all about communication and compromise.

CHAPTER 7

Underrepresented Groups

Just as women have faced discrimination in academics, so have members of underrepresented groups. If you feel you are experiencing discrimination because you are a member of such a group, you should discuss this with your mentors and decide upon a course of action. If your mentors do not feel that this is a case of discrimination, you need to reevaluate the situation. Everyone has disappointments. Those who succeed have been able to pick themselves up, dust themselves off, and continue with more resolve than ever.

A POSITIVE ATTITUDE GOES A LONG WAY

We all need to appreciate ourselves for our good qualities, know our limitations, and accept responsibility for our behavior. We all need help at times, and in accepting that help, we acknowledge the need to help others in similar circumstances. It is important to remember that achieving our goals requires a great deal of hard work. We have all known individuals for whom everything seems to come easily. Eventually even they reach a point when they have to work hard to achieve their goals.

Unfortunately, there are still some individuals who harbor prejudice against members of groups other than their own. When they see a member of another group, they do not see the qualities of that individual; instead they see the negative stereotype they attribute to members of that group. It is not clear whether anyone will be able to change their perspective; however, that does not mean that you have to be one of their victims. If you encounter such an individual, you need to consult your mentors regarding the best strategy to deal with them.

BENEFITS OF SUPPORT GROUPS FOR UNDERREPRESENTED MEMBERS

Certainly it is inspiring to see members of your group who have achieved goals you share. Their stories are compelling and can provide examples of strategies that you may be able to use. It is important to know that you have resources to turn to for advice and support.

Such groups may also offer training in the unwritten aspects of your career: presenting, publishing, and securing grant support for your work; establishing a research network; teaching; pursuing leadership opportunities; achieving promotion and tenure; and balancing your personal and professional lives.

PROBLEMS WITH SUPPORT PROGRAMS FOR UNDERREPRESENTED GROUPS

Unfortunately, some support programs operate with the unwritten assumption that group membership means lack of skill in one or more areas. They may subject all group members to the same remedial program regardless of skill. This is simply another example of prejudice.

Summer Program for Research

Sam is a member of an underrepresented group. Sam is an excellent undergraduate student. Sam's professor describes Sam as the best biochemistry student in thirty years. Sam applies to a summer research program designed to recruit prospective graduate students to a well-known medical school. For the first time, Sam indicates membership in an underrepresented group on the application. Instead of the summer research program Sam applied for, Sam is assigned a remedial program. Sam is very upset. Sam's professor contacts a colleague at the medical school who is able to have Sam switched to the research program. Fortunately, Sam attends the research program and finds it very rewarding. Sam goes on to earn an MD and a PhD, but never considers applying to that particular medical school for the MD/PhD program, a residency, a fellowship, or a faculty position.

ASSUMPTIONS REGARDING A RETURN TO THE UNDERREPRESENTED GROUP

Sometimes, explicitly or implicitly, members of underrepresented groups are expected to return to their community to provide service. However, there are many ways to "pay back" your community. Throughout your training and career you can provide mentoring to other members of your group and of other underrepresented groups. You can assist with recruitment during training and as an academic professional. You can assist with K–12 pipeline programs.

You need to be careful that these pay-back activities do not dominate your professional time. You could be so overcommitted to helping others that you don't have time to pursue your own career goals adequately. You need to be selective and focus on activities that are a meaningful and effective use of your time.

Be Selective When Choosing Administrative Responsibilities

Kim was a new assistant professor. As a member of an underrepresented group, Kim was asked to serve on the admissions committee. This would consume 50 percent of Kim's time for six months of the year. In addition, Kim was asked to serve on all of the department's faculty recruitment committees. The chancellor asked Kim to be a member of the university's diversity committee. The National Institutes of Health wanted Kim to join a Study Section. The national professional organization in Kim's field wanted Kim to serve on an important committee. While considering these opportunities, Kim consulted a mentor. The mentor reminded Kim that, as a new assistant professor, Kim's first commitment was to be successful in obtaining promotion and tenure. The mentor suggested that Kim select the one most important opportunity and pursue that. Kim decided to focus on the professional organization's committee. Kim will have plenty of time to pursue the other opportunities after achieving promotion and tenure.

CHAPTER 8

Preparing for Postgraduate or Postprofessional Training

Certainly many graduate and professional degrees lead directly to a position without further training or with a limited amount of additional training. Graduate training resulting in a master's degree (e.g., master of arts, masters in business administration, masters of engineering, master of fine arts, masters in education, masters of public health, masters of public policy, master of science) and doctorate can lead directly to jobs and career paths. Professional degrees such as doctor of dental science, doctor of medicine, or doctor of veterinary medicine may require a year of internship before licensure. Because licensure in these areas is regulated at the individual state level, the rules can vary significantly.

However, if you want to learn a new method, explore a new field, acquire credentials for specialty or subspecialty board certification, or work at the interface of two disciplines, you will want to pursue further training. You may find fewer training sites to choose from compared to when you were selecting your initial training program. You may also find more intense competition for these more limited positions. Just as you carefully researched graduate and professional schools, you need to evaluate each opportunity carefully and make sure that you meet the criteria for selection.

POSTDOCTORAL TRAINING

The purpose of postdoctoral training is to enhance your marketability for a job. As with selecting a training program, you will want to select a good environment, but the most important consideration is a mentor committed to furthering your career and helping you to take the next step. It is also important to have financial support for your fellowship training. Having a supportive infrastructure for fellowship training within the department, center, or institute is an added bonus. You will benefit from the larger group of mentors, the interactions with other postdoctoral fellows, and formal courses to supplement your graduate training, particularly if you are embarking on a new area of investigation.

After your fellowship, you will have to have proven skills to earn a salary. The skills required for your position may involve teaching, research (supported by grants, patents, and/or a job in industry), community service, and/or administration.

As a postdoctoral fellow, you have numerous opportunities to get teaching experience. You should take advantage of every opportunity to present talks at group meetings, seminars, and professional meetings. There might also be the possibility of serving as a teaching assistant. You could organize and lead a course or a journal club. To improve your teaching and to validate your teaching skills, you should ask students, peers, and faculty members to evaluate your teaching, even if this is not a routine process at your institution.

You may also have the opportunity to demonstrate your potential for independent research and gain experience in preparing grant applications. You should apply for an individual postdoctoral fellowship award to establish your own track record. You should apply for career development awards from private foundations and/or federal agencies that provide transition from postdoctoral training to faculty position, since such funding would be quite appealing to departments hiring new faculty members. You should take advantage of any opportunity to develop a patentable product, especially if you are interested in industry.

To develop your administrative skills you can volunteer to serve as the postdoctoral committee member for academic, industrial, or professional organizations. You can be active in postdoctoral trainee organizations.

You might consider further training as a postdoctoral fellow to differentiate yourself from the typical job applicant. If you are interested in working in policy, you might consider law school, a PhD in public policy, a fellowship in a cross-disciplinary center with a policy focus, or an internship with a federal, state, or local government agency. If you want to develop and test new therapies for patients (clinical trials research), you might consider an MS in clinical research or membership on an institutional review board that reviews human subjects research protocols. If you would like a position in biotechnology, you could do a fellowship in industry or at an engineering school, or obtain a master's in business administration or a master's in biostatistics. If you have a PhD and would like to direct a clinical genetics laboratory, you might consider a fellowship that prepares you for certification by the American Board of Medical Genetics.

YOU NEED TO THINK BEYOND THE TYPICAL DEPARTMENT IN A LARGE UNIVERSITY OR PROFESSIONAL SCHOOL

The reality is that at this time, approximately only 20 percent of PhDs find tenure-track academic positions in large universities or professional schools. However, other paths provide satisfying academic careers. There are tenure-track positions at small colleges that require a focus on teaching, and adjunct faculty appointments without tenure in large universities or professional schools. Private companies can be very academic, encouraging publication, grant applications, and affiliation with professional organizations. They may even be affiliated with a large university or professional school. Federal or state laboratories provide opportunities for research, training, and administration. Publishers provide opportunities for authors who wish to pursue independent writing careers and for in-house editors for professional journals, textbooks, and trade books. Professional organizations have a wide variety of administrative opportunities.

Administrative opportunities in large universities or professional schools include student services, instructional services, research services (e.g., an institutional review board for human subjects research or an animal research committee for animal subjects research), and professional services (e.g., administration of clinical laboratories). Administrative opportunities in private companies include human resources, marketing, and interfacing with regulatory agencies. The National Institutes of Health has a variety of administrative positions including program officers in the Centers and Institutes and scientific review administrators for the Study Sections. Other governmental agencies with doctoral-level administrators include the Centers for Disease Control, the Food and Drug Administration, and the Patent and Trademark Office.

Serving as a consultant in the following areas can also be rewarding: research methodology and statistics; writing manuscripts and grants; laboratory design; drug development; clinical laboratory testing; and medical-legal care analysis and expert witness testimony. You can work independently or be part of a larger company—for example, a national or international consulting firm.

You have a long career ahead of you. There are many opportunities to reinvent yourself, to continue to learn, and to develop your potential. Postdoctoral fellowships may be the first time that you do not have a prescribed schedule of courses and committee meetings to attend. You need to establish long- and short-term goals for yourself with your mentor. You and your mentor then need to review these goals formally in a scheduled meeting at least every six months to be sure you are making progress.

The Goals of the Mentor and Mentee May Conflict

Lee is extremely pleased to obtain a postdoctoral fellowship with Dr. Johnson, a very visible and productive scientist. Lee will learn an exciting new research method as part of the training program. Lee is planning to spend two to three years with Dr. Johnson before applying for a job. Once Lee has joined the group, it becomes clear that the two- to three-year plan does not coincide with Dr. Johnson's insistence that any paper from the group must be published in a top-tier journal. These journals typically require about five to seven years of work. Lee's choices are to stay much longer than originally planned or to settle for eventual middle authorship on the two to three years of work that will be involved to start the project. Neither of these possibilities will help Lee move on to a tenure-track academic job in the near future. Lee begins to think about alternative career paths beyond universities.

POSTPROFESSIONAL TRAINING

With a professional degree, you may need to complete an internship, study for boards and/or the bar, and pass the boards and/or the bar before licensure. If you are interested in a particular specialty or subspecialty, you may have been preparing for this during professional school by selecting appropriate electives and/or seeking additional experience beyond the routine requirements of the training program. If you are interested in a particular institution for specialty or subspecialty training, taking an elective at this institution provides you the opportunity to determine if it is a good fit for you and gives the faculty the chance to see if you have the qualities valued by the group. The recommendations you receive from those at the institution will mean more than those from your training program.

An Unusual Combination of Degrees
May Provide Unique Preparation

Kelly completed law school and realized he needed more training to further his career in health care policy. He decided to get an MD and complete a residency so he could fix health care from the inside. This combination of degrees and his experience provided a strong basis for his career. Kelly realized that he had the ability to think like a lawyer and a physician's knowledge of the health care system.

Remember that life is a marathon, not a hundred-yard dash. You can always return to specialty or subspecialty training after years of practice. Applying to these training programs is more competitive than admission to professional school, but your experience will make you a more attractive applicant.

The purpose of your specialty/subspecialty training is to provide you with the credentials and the experience to pursue certification and positions in this area. Before applying, visit the accrediting office's website and determine which programs are accredited. Interviewing offers the opportunity to determine if the setting will provide you with the number and quality of experiences you are seeking. You will also get a feel for the current trainees and the faculty, as well as the type of training environment they provide. A supportive, positive program that is interested in your future career is ideal. You also need to determine which career paths are open to those completing the program. Be sure that the program will prepare you to pursue the mix of professional, teaching, research, and administrative activities that interests you.

Promises Should Be Kept in Subspecialty Training

Kim wants to be an adult cardiologist who studies the effectiveness of medications for heart failure. After completing medical school and a residency in internal medicine, Kim moves to a cardiology fellowship at a prestigious university with a well-regarded program in adult cardiology. While interviewing, Kim was told that there would be ample opportunity for research with Dr. Anderson, a noted clinical researcher in heart failure. Dr. Anderson assured Kim that the fellowship program would guarantee 75 percent time for research in the second and third years, as required by the training grant that would be supporting Kim. The first year of fellowship was very intense clinically. During the second year, Kim did not have any more time for research. A faculty member left unexpectedly, and recruiting a replacement was slow-going; a third-year fellow was pregnant with twins and on bed rest; a second-year fellow could not handle the stress and left the program; and a first-year fellow decided not to come when his partner was transferred to another city. Kim realized that relief from clinical work and time for research was not going to happen in the foreseeable future. Kim discussed these issues with Dr. Anderson, who worked with the division chief to hire two cardiologists in private practice to help with the patient load. Kim was then able to have time for research, in accord with the training grant and Kim's goals. Kim understood why other academic cardiology training programs had indicated that they recommended a commitment of five years in fellowship training to be adequately prepared.

SET GOALS DURING YOUR
POSTPROFESSIONAL TRAINING

You need to be sure that you meet all the requirements for your specialty or subspecialty certification. You should meet every six months with the director of the program to be sure you are meeting the deadlines and making progress toward your goals. In addition, you should meet every six months with your mentor to set short- and long-term goals and assess your progress. You are laying the basis for your career at this time, so you must approach this period of your training in a very structured manner.

If you see yourself teaching in the future, you need to seek experiences to give lectures to trainees, participate in journal clubs, present your research findings, and participate in professional meetings. If you see yourself doing research, you need to invest the time and energy required to pursue your research project. This may require taking courses or even pursuing another degree. If you see yourself in an administrative role, you need to seek leadership responsibilities as part of your training program (e.g., as a trainee member of your specialty/subspecialty residency review committee) and in professional organizations (e.g., as a member of the trainee section).

You Need to Work as Hard at Research as at Other Aspects of Your Career

Kelly's subspecialty training required a great deal of clinical activity. This work was intensive, demanding, and tiring. When she was supposed to be doing her research, she used this time to catch up on sleep, exercise, social relationships, and traveling. She would begin her research between 10 and 11 A.M. and leave by 3 P.M. because of social commitments and the gym. During each research block, she took a personal trip out of town for at least a long weekend and sometimes for a week or more. She avoided group research meetings and individual meetings with her mentor. After about six months, her research mentor asked Kelly into her office. They discussed Kelly's lack of commitment to research. Her mentor reminded Kelly that it took quite a number of years to acquire her clinical skills. If Kelly wanted to succeed in research, she needed to put in sufficient time and energy.

CHAPTER 9

Applying for, Negotiating, and Choosing a Job

USE YOUR TRAINING PROGRAMS
TO PREPARE FOR YOUR JOB

Develop your research background through professional meeting presentations, papers, grant proposals, and intellectual property. Develop your professional pathway and prepare for relevant examinations, such as specialty and subspecialty boards. Develop your teaching skills and collect trainee and peer evaluations to use as evidence of your teaching ability. Consider pursuing other degrees (e.g., MD, PhD, JD, MPH, MS, MBA) to match your professional pathway goals.

YOUR CURRICULUM VITAE IS A REFLECTION
OF YOUR PROFESSIONAL CAREER

Your first contact with a potential mentor or department chair is often through your curriculum vitae (CV). Daily maintenance of your CV enables you to store historical information about your professional career for a variety of purposes. Your CV contains the biographical information you need when you apply for a grant, give a talk, apply for promotion and tenure, contribute your information to the department's website, apply for committee membership for a professional organization, or teach a course for continuing education credit. In addition to providing an opportunity to make a good first impression, your CV allows storage of information that you frequently need on short notice. A well-organized CV offers an excellent impression of your organizational skills.

Your Curriculum Vitae Is Not a Business Résumé

This document should have the heading "Curriculum Vitae." Technically, your CV refers only to your professional history, with the bibliography being a separate document. By common practice, however, both are usually embodied in a single CV document. A CV is not a business résumé. Whereas business résumés are organized from most current to oldest item in each category and are relatively short (usually one or two pages in length), CVs should be organized from oldest to most recent in each category and should be as long as they need to be to include all of your professional activities.

Include Your Name, Titles, "Coordinates," and as Much Personal Information As Is Comfortable for You

The first piece of information in your CV is your name, along with your graduate degree(s) and your current position. Be sure to list all of your professional titles, including membership in research units and centers or institutes outside your departmental affiliations. You should then provide your current address, phone, fax, e-mail address, and citizenship or immigration status. Optional information includes your birthplace, birth date, marital status, length of marriage, significant other's name, significant other's occupation, and children's names and ages.

Your Educational Information Should Provide an Accurate and Complete Review of Your Professional Preparation

Information regarding your education and training should include the time period, discipline, institution, city, state/province, country, degree, and date of degree. Description of your professional experience should include the time period from starting to ending year, your title, department, institution/company/agency, city, state/province, and country. If you took time off or had breaks in your education or professional experience, consider explaining these briefly. Speculation about the reasons for these holes in your career may be damaging to you as a trainee or job candidate.

Your Honors Represent Important Distinguishing Characteristics

Under "Honors" you should include election to any honor societies, graduation with honors, fellowships, scholarships, awards, and named or memorial lectures you have presented. You should include the year the honor was received, as well as the location.

Committee Memberships Show Good Citizenship and Leadership

You should specify your membership in or chairmanship of any committees outside your employing institution. These committees may include professional organizations, volunteer organizations, and grant review panels. Listings should include the time period, your role, the name of the committee, the organization, and any other pertinent information.

Organize Your Activities by Institution

For each institution/company/agency where you have had an appointment, indicate your teaching, committee, and professional activities. Each of these entries should include the time period during which you were involved, your role, and the name of the activity. Under teaching, you should include lectures you presented, courses you organized, curricula you developed, and training programs in which you were involved. Under committees, list the committee, the duration of your involvement, and your role. Under professional activities, you should detail the duration, the nature of these activities, and your role.

Include Any Professional Licenses or Certificates

Indicate in your CV your professional certifications (such as specialty and subspecialty board certificates), the dates they were obtained, and any identification numbers. You should also include your professional licenses, the dates when they were obtained, and the identification numbers. Some jobs you apply for may require certification or licensure.

If you have a time-limited certificate, be sure to include whether or not your certification is current. If a license has expired, be sure to include the ending date. The status of certificates and licenses is easy for a potential employer to ascertain. Not including this information, even only through an oversight, might be perceived as attempted deception on your part, and you might not be considered a suitable candidate for the position.

List Your Grants and Students/Trainees

List all of your successful and currently pending grant proposals, providing duration, title, total award amount, and your role. If there are grants that you were awarded and could not accept—for example, because of duplication of awards—you should include those and list them as awarded but declined with a brief statement of the reason.

List all of your trainees, their training duration, their research projects, dates of degrees, and training institutions. You may also wish to include awards your mentees received and their current positions and institutions.

Separate Your Peer-Reviewed Publications from Other Categories

In your bibliography, you should include several categories of publications. Each of the following should have a separate section and should be listed chronologically from oldest to most recent: peer-reviewed publications; reviews and chapters; books; and other publications, such as letters to editors and symposia presentations. Published abstracts should follow. The last section should include your invited oral presentations with the title of the presentation, type of talk (e.g., grand rounds, research seminar), department, institution, city, state/province, country, and date. If the oral presentation was at a professional meeting, you should include the title of the presentation, meeting, city, state/province, country, and date.

Your Curriculum Vitae Is a Repository of Valuable Information, So Keep It Up-to-Date

As your career progresses, it will become increasingly difficult to recover lost information. You will be wise to keep your CV updated on a daily basis, as new information becomes available. Do not put this off, or information will be lost. Keeping this document current will facilitate your ability to respond to requests for career information stored in your CV. It is disconcerting to receive someone's CV supporting their job candidacy, only to realize it is one or more years out of date. Such delinquency raises serious concerns about the person's organizational skills and interest in the position.

SELECTING A POSITION

How Do You Find Out What Jobs Are Available?

Good sources for potential positions include your colleagues, mentors, division chief, and department chair. You should consult professional journals and look for job fairs at professional meetings. However, the best positions are often obtained by word-of-mouth and by nomination on the part of someone you know who also knows the individuals who are hiring.

Never Consider a Lateral Move except . . .

Once you have completed your training, you will be applying for a position. If you already have a job, you may be considering a move to a different institution, company, or agency. Essentially the same criteria apply to both situations. If you already have a position, you should not waste your time and energy applying for a job that would be a lateral move, unless circumstances at your current place make staying there professionally untenable. Valid reasons for considering a lateral move include power or gender abuse, lack of resources for professional development, no possibility of promotion, or impossible demands for teaching or service that interfere with professional development.

If You Don't Stand Up for Yourself, Others Can Take Advantage of You

Stacy was pleased to be offered a faculty position at the end of fellowship training. When no new fellows were recruited, Stacy asked if there would be any in the future and was told that there was no funding for fellowship training. Stacy was very busy clinically. Stacy was on call 24/7. When Stacy left town, even if it was an alleged vacation, the staff would call with questions regarding patient management. Stacy was responsible for obtaining grants to support salary and research. There were no new faculty or fellows. Stacy would always be the most junior person in the division, and the perpetual "fellow," even as an assistant professor. One day Stacy saw a listing of national average salaries for medical school faculty. For someone at Stacy's level and subspecialty, Stacy's salary was at the tenth percentile. It was time to move on. Stacy secured a position at another institution that involved a promotion to associate professor with tenure, the position of division chief, fellows and other faculty members to share the clinical load, an exciting research environment, and a significant increase in salary.

Job Searching Takes a Lot of Energy

You must recognize that looking for a job takes time away from your professional development. Job searches consume your time with preparing for interviews and job talks, traveling, enduring several days of interviews, and providing the requested written response after the visit regarding your plans and your needs. You should consider moving if it means a promotion, tenure, resources to support your work, a more prestigious institution, better trainees, and/or the opportunity to improve your career through better interactions with new colleagues. Unfortunately, it may take an offer from another institution to inspire appreciation of you in your current position.

Do not get the reputation, however, of someone who is always look-ing just to improve your situation at your home institution. Individuals who have such a reputation often find that, when they need to make a move, no other institution is interested in wasting time interviewing them.

You Should Inform Your Current Chair
If You Are Looking at Another Job

Chris was flattered when Dr. Smith, the chair of a prestigious depart-ment at a different university, took an interest in her presentation at a national professional meeting. After discussing Chris's research, Dr. Smith suggested that Chris apply for an opening for an assistant profes-sorship in Dr. Smith's department. Chris was already an assistant pro-fessor at a equally prestigious institution, but the attention from Dr. Smith was flattering. Chris had been frustrated recently when a new assistant professor was hired with a better research package than Chris had received a year earlier. Chris was also concerned that she was carrying a heavier teaching load than the senior professors in her department and that there was not sufficient office staff to process her manuscripts and grant proposals. Chris submitted her CV to Dr. Smith and visited his department. Chris did not tell her department chair, Dr. Jones, about her interest in another job. She simply said she was going to give a sem-inar. Chris's interview went well, and Dr. Smith offered Chris the posi-tion. Chris was attracted to the position because Dr. Smith offered a large research package, minimal teaching responsibilities, and access to more office staff support. When Chris told Dr. Jones that she was resigning to take a position with Dr. Smith, Dr. Jones was surprised and upset that Chris would consider another job without first telling him. Dr. Jones offered more research support, a reduced teaching load, and increased access to office staff. Dr. Jones suggested that Chris consider the impact of a move on her research program. Closing her current research pro-gram, relocating, and hiring new staff would put Chris at least six months behind. Considering the career and personal costs of moving, Chris declined Dr. Smith's offer and decided to stay in Dr. Jones's department.

Beware of Becoming an Academic Itinerant

Even if the institution you are moving to pays your complete moving costs, you will still bear the costs of setting up a household in a new location, your work will be interrupted, and moving will impact your trainees, staff, colleagues, and family members. The professional disruption can reduce productivity and interfere with your ability to compete with your colleagues for grants.

We all know academic itinerants who move every three to five years without any obvious change in status or apparent reason for moving. One wonders whether they would have been more productive had they not moved so often. One questions the reasons for all of the moves. Speculation often includes the attention paid to them during the recruiting process, a desire to have the funding from recruiting packages, a difficulty getting along with colleagues after more than a few years, and other unflattering explanations. A reputation like this will become well known nationally and will impact grant and manuscript reviews.

Demand Protection to Pursue the Career Path
Agreed Upon When You Were Hired

Sam completed training and became a faculty member at the same institution. While he had all the responsibilities of a new faculty member, he still retained the heavy service commitment of a trainee. After a year of frustration and lack of progress, Sam met with his department chair to discuss the situation. The chair agreed that Sam would be hard-pressed to meet the criteria for promotion and tenure. The chair promised to begin recruiting a new trainee right away and to have one in place in less than a year. Twelve months later, there had been no attempt to recruit a trainee and Sam was still in an untenable position. He felt a great deal of loyalty to the department and the chair, but did not want to fail as a faculty member because of unrealistic expectations. Sam went to the chair and said that he had to begin a search for a position in another institution that would recognize his faculty status and provide trainees to help with the service workload. Sam's job search jolted the chair into reality. The chair immediately divided the service workload equally among all of the faculty members in the department and submitted advertisements to the leading professional journals to recruit not one, but two trainees. The chair's willingness to protect Sam's time and the prospect of trainees convinced Sam to forego a job search.

Criteria for Evaluating a Position

You should use the following criteria to evaluate a position:

· Commitment of the department to the career development of junior members

· Sufficient resources to support the professional activities of junior members, thereby providing the basis for their future success and independence

· Appropriate expectations regarding time allotted for teaching, administration, and/or service, so new members have time for academic development

· Availability of dedicated mentors to support the development of professional skills and to provide understanding of the criteria for promotion

· A larger community with similar interests, both within and beyond the department, to provide the opportunity for exchange of ideas and professional resources, and the potential for collaborative efforts

· Access to work with trainees at all levels

· Integrity of the department head: Will the offer letter be honored?

· Financial considerations, including salary, benefits, and support for professional development

· Lifestyle: Is this a place where you feel comfortable and can enjoy your time away from your professional life?

The Contract Is Only as Good as the Two People Signing It

Kim carefully negotiated a wonderful start-up package for his first faculty position. He was offered sufficient funds for his research, enough space for an office and laboratory, and a guarantee of 80 percent protected time for research. When Kim arrived, the funds and space were available as promised. However, instead of spending 20 percent of his time on service activities, Kim was spending 80 percent. When Kim spoke to other junior faculty, they all said that the chair had not honored significant portions of their offer letters. When Kim spoke to the chair, the chair said that the letter was written before the unanticipated departure of two senior faculty members and the offer letter was not binding anyway. Kim took his case to the dean. The dean invited the chair to join his meeting with Kim. The dean suggested that the chair immediately hire two temporary faculty members and replace the two former faculty members as soon as possible. The dean stated that once the two temporary faculty members were on board, Kim should spend 20 percent of his time on service commitments. In addition, the dean would meet with all other junior faculty members in the department to ensure that the chair met the promises in their offer letters as well.

Any Job Requires That You Earn Your Salary

Every job requires that you bring in a combination of research, professional service, teaching, and administrative dollars. You need to determine the combination of support that works best for you, and decide whether the requirements of the position will allow you to earn your salary. If you are concerned that the job will not allow you to achieve the financial requirements of the position, you need to address these concerns before you accept the position.

The Interviewee Is Also an Interviewer

To obtain information about a department, you need to ask questions when you visit. You should clearly define the qualities of a department that are important to you and determine whether this department has these qualities, particularly the core features that you feel are absolutely required for your success. You should carefully read all written material the department provides to determine whether the information is consistent and whether the institution/company/agency shares your values. You should visit the department and the institution/company/agency webpages to check the information and the image projected. You should perform a literature search for publications produced by the department and the institution/company/agency, particularly by anyone with whom you will interact during your visit. Such information will indicate to the interviewers your sincere interest in the position and will assist you in conversations during your formal interviews, as well as during more casual interactions, such as at meals. You should talk to colleagues with previous connections to the institution/company/agency, since they frequently can give you inside information, but you must also be aware of their biases and the timeliness of their data. Why did they leave? When did they leave? Have they maintained contact? You should ask colleagues who have no past history with the institution/company/agency to give their impression of the status, supportiveness, and mentoring of the environment.

Evaluate a Program's Trajectory as Well as Its Current and Historical Position

Lee received two job offers for her first faculty appointment. One was from a well-established, prestigious private university of long standing.

The department had produced a long line of famous investigators over the years. The second position was with a comparatively new state university. This department was growing, and a number of new faculty members were recognized as representing the future of the discipline. When Lee compared the two offers, the private university offered a lower salary, smaller research funding, and heavier teaching load than the state university. Still, Lee was leaning toward the private university based on prestige and history. Fortunately, Lee compared the NIH rankings of research grant funding in the past ten years. These rankings showed that the private university had decreasing money each year while the state university had increasing funding. Lee realized if current trends continued, within two years the state university would have more NIH funding than the private university. Lee's mentor, Dr. Smith, suggested that the dean at the private university was more interested in creating administrative infrastructure than in developing research programs. The dean at the state university had a widely publicized goal of being in the top ten state universities in terms of NIH funding. With Dr. Smith's advice, Lee accepted the position at the state university.

Ask to Meet with a Broad Range of Individuals

Sometimes you will be asked to suggest individuals with whom you would like to meet on your visit. Because of your research, you should have some ideas regarding colleagues in your area of interest, especially those directing the training program, a relevant center or program, or a core facility. You should also suggest meeting with trainees, recent recruits, the department chair, the division head, and the director of education. Go on the Internet and identify a list of individuals with whom you want to meet. If you respond to this request by saying that you will leave the schedule up to your hosts, they may speculate that you are lazy or do not care about this position.

Broad Input Will Allow You to Formulate a Plan

Sandy and Sam were the two finalists for the position of department chair at an outstanding institution. Before their second visit, both were asked whom they would like to meet with on their second visit. Sandy wanted to learn as much as possible about the department and the institution. He asked to meet with those he knew in his area of research. He also did a literature search to determine other potential research collaborators. Sandy also contacted several friends who knew the department and institution and queried them for the names of important individuals in the research and administrative hierarchy. In addition to these people, Sandy also asked to meet with trainees and recent additions to the department. Sam was more concerned about living conditions and salary issues. Sam did not ask to meet with anyone. At the end of their visits, the dean asked both Sandy and Sam to formulate a plan for the department. Sandy felt well prepared to do so, while Sam did not know where to begin. The dean was very impressed with Sandy's understanding of the department and the university. Sandy was named to be the department chair.

Your Search Should Be Data Driven

You should also request information about the national standing of the department and the institution regarding training, grant and philanthropic dollars, quality of professional activities, membership of individuals in prestigious societies, and professional awards to individuals. In addition to current data, you should review past levels to determine if the department or school is on the way up or on the way down in standings. If you will be a junior faculty member, it is more important to determine where the institution and department will be in the next five to ten years, rather than where it is today. You need to evaluate trends, not just current status. Per capita rankings are most helpful. You should not be satisfied with self-serving, unsupported propaganda, and you should ask to see the data. It is better to be at a lesser institution that is on the way up than at a better place that is on the way down.

Success Requires Resources and Creativity

Lee was thrilled to be named the division chief in a prestigious department at a renowned university. Once Lee arrived, it became painfully obvious that this administrative position, rather than stimulating his career, would probably end it. As division head, Lee was responsible for a training program for which there were no funds from the department and no staff support. The only way to support such a program was through service income. Before taking the position, it had never occurred to Lee to question why none of the current faculty wanted to become the division head. Now Lee understood. Fortunately, Lee had experience with a little-used federal funding program and knew of a niche group of foundations in his discipline. By pitching proposals to this funding network, Lee was able to develop a new resource base for the division.

THE ONE-HOUR TALK, INCLUDING
THE JOB APPLICATION SEMINAR

The One-Hour Talk Is Really Forty
to Forty-five Minutes Long

First, consider the length of the one-hour talk. It is not one hour long. It will start five to ten minutes late, and you should leave at least ten minutes for questions at the end. You should also recognize that different disciplines have different conventions, and different countries have different expectations. If you do not know the culture, you should ask your host how long your presentation will be when you are scheduling your talk.

Determine the Expectations of Your Hosts

Dr. Johnson was pleased to be invited to talk at three different major institutions in a foreign city. Dr. Johnson reasoned that some of the same individuals would attend all three lectures. Dr. Johnson developed talks on three different aspects of the same topic, but didn't check with the host until the evening before the first lecture. Dr. Johnson did not believe it when the host said it was expected that Dr. Johnson would talk for three hours. Unfortunately the most important talk was the next day. Dr. Johnson gave a one-hour talk very slowly. The host was dismayed that the talk was so short. For the next talks, Dr. Johnson combined all three talks and talked for three hours. This was much appreciated by the host.

Know the Audience for Your Presentation

Consider the venue as you prepare your talk. You should ask your host about the audience and the type of talk they are expecting. Is it scientific or clinical? Will it be an educational or peer presentation? Are the members of the audience from your specialty or subspecialty, or will they be a more general group? Will they be a homogeneous group, or a group with mixed backgrounds and knowledge bases? Is this a local, regional, national, or international meeting?

You need to consider the sophistication of your audience and be sure that you define any terms that may be unfamiliar to them. Avoid jargon. Another variable is the audience's level of interest. Are they attending the meeting for educational purposes, intellectual curiosity, or recreational purposes? If it is a meeting in a destination location, you may have to be particularly creative with your title to get the audience to attend your talk, and bold in your presentation to keep them. You need to determine if your talk is a lecture, seminar, or problem-solving session.

Tailor Your Talk to Your Audience and Time Constraints

Kim always gives the same talk in every setting. The one-hour talk is a very sophisticated presentation of his research, appropriate for his colleagues in that field. When he is invited to present a ten-minute talk at a professional meeting, he uses essentially the same PowerPoint presentation and stops in the middle when the session chair tells him time has run out. When he is asked to give a half-hour talk to the public, he rushes through the same presentation and almost finishes. When he is giving a one-hour lecture to students, he uses the same talk and never bothers to define terms. Eventually, invitations to present talks at professional meetings and other institutions/companies/agencies decrease. Kim's lack of enthusiasm for his audience, as seen in his lack of interest in developing an appropriate presentation, led to a lack of enthusiasm for inviting Kim to speak.

Your Title Should Be Interesting, Your Organization Clear, and Any Potential Conflicts Apparent

As you prepare your talk, you need to select a title that will interest your potential audience and will attract them to your presentation. The title of your last paper is probably too long. You need to acknowledge the source of support for your research. If a representative of the grant-funding agency that supports your research is in the audience, they would be concerned if you did not indicate their support for your work. You need to state any potential conflicts of interest. For example, if a drug company is supporting your clinical research on its product, you need to communicate this to your audience and follow the host group's rules to resolve this conflict of interest. For example, you may be able to use only peer-reviewed published data to support your conclusions and recommendations.

When you are invited to speak, you should ask if you need to provide an outline, learning objectives, and/or a manuscript. Some organizers feel that you are less likely to accept an invitation if you will be required to provide a manuscript. These organizers don't let you know ahead of time and spring their request on you just before your presentation. At that point, you are committed to give your talk. It's far better for you to prevent this scenario by asking before you agree to give the talk.

Be Cautious about Relinquishing the Copyright to Your Presentation

If your presentation will be published in any form, including on the Internet, you may no longer own the copyright to the material. You do not want the publication of your presentation to include any new data that you are planning to publish in a peer-reviewed journal. If that should occur, you will be unable to publish this new material in an original article since it has already been published. If copies of your PowerPoint presentation are published, your slides may be owned by the publisher if they copyright the publication. Using your PowerPoint program in future presentations may be a violation of this copyright.

Organize Your Presentation like a Banquet

As you plan your talk, think about the last banquet you attended. Think of the introduction and summary as the appetizer and dessert. In between, you can accommodate no more than three to five courses. If you eat more than that, you would be unable to remember what you ate. If you have more than three to five points in a one-hour talk, your audience will be unable to remember them. For shorter presentations, you will need to pare down the number of points you make.

Plan Your Presentation Like a Script or Screenplay

Remember what maintains an audience's interest in a play or movie. Plan your talk to include changes of pace. If you can organize your talk to create suspense, you should do this. For example, if you have been successful in answering a long-standing question in the field, solving a mystery as it were, then build to that as the climax. You might include the occasional wrong turn that you made, if you can do so without leading to confusion. The audience will bond to someone who, like themselves, is imperfect.

Preparation Reduces Anxiety

To deal with your anxiety, you should practice in front of your colleagues, who will ask you difficult questions. Your colleagues will make sure that your PowerPoint program is correct, your organization is reasonable, and that your presentation fits in the time allotted.

Remember to Breathe

You need to remember to breathe when you are speaking. At the time of your presentation, you may want to have a glass of water handy. Taking a sip of water will help you to catch your breath if you are nervous. Select several individuals in the audience and make eye contact with them. The audience does not expect you to make eye contact with each of them, but they do not want to see you scanning the audience and not looking at any individuals. Making eye contact, even though only with a limited number of individuals, will help the audience bond with you, and it makes a large audience seem smaller.

For many nervous speakers, the first three to five minutes are the most difficult. You should visualize your presentations that have gone well and recognize that most of your fears are not based on any prior experiences you have had as a speaker.

Enjoy Yourself and Your Audience Will Enjoy Your Talk

Your practice with your colleagues should give you confidence and help you relax while you talk. Enjoy yourself, maintaining a sense of humor without trying to be a comedian. Know your topic thoroughly and be friendly and enthusiastic.

Remember to Say Thank You So Your Audience Knows When to Clap

At the end of your presentation you should say "thank you." You have been at presentations where the audience does not know if it is over. A pleasant and sincere thank-you is a polite way of letting your audience know the presentation has concluded and they can applaud.

It's Not Over until It's Over

After the applause, do not relax, because it is not over. You still have the questions and answers.

Be brief when answering questions. Your responses should be no more than one to three sentences in length. Be sure to repeat the question if the audience cannot hear the questioner. This will also allow you time to plan your answer.

Whenever possible, you should attend the entire session to hear other presentations. After all, you wanted an attentive audience for your talk. At the end of the session, you should thank the moderator and your host.

Experience Gives You Confidence

Chris was terrified of speaking in public. As a graduate student, he would have to make numerous presentations to his committee in addition to presentations at scientific meetings. Chris's mentor, Dr. Smith, was very sensitive to his fears. Dr. Smith realized how important presentations would be to Chris's success as a graduate student and throughout his academic career. Dr. Smith always arranged weekly lab meetings so that each member of the group presented their progress, problems, and plans at every meeting. Chris, like everyone else, gave a weekly presentation of his work, gave a half hour progress report every six months, and presented a journal article every six months. In addition, Dr. Smith would ask Chris to explain a method or explain an alternative research design at the weekly group meeting. To further desensitize Chris, Dr. Smith had him submit abstracts for regional professional meetings where there was a good chance of Chris giving a platform presentation. Dr. Smith made sure Chris was well practiced before he had to give a one-hour seminar as part of his job interview. Chris soon gained confidence in his abilities, became an excellent speaker, and was ready for his job talk.

THE JOB INTERVIEW

Prepare Yourself to Negotiate from a Position of Strength

Think about your negotiation from the point of view of your possible future supervisor. What can you do that helps to meet his or her needs? There are three things that no one can dispute: your degrees; money you may bring to the institution/company/agency (grants, donors); and publications.

Your training may make you a hot commodity. Whether it's your subspecialty or your degree, or your combination of degrees and training, you still need to look for the opportunity that will allow you to develop your career in the short term and in the long term.

Money is always an issue. There is never enough. If you come with a funded grant, if you have a history of generating patents, or if you have appeal to a potential donor, you have a stronger negotiating position than someone who doesn't have these potential sources. You have already demonstrated your skills in the area. You are a known quantity.

Academic Recruiters Are Not
Human Resources Professionals

Sam went for an interview for a faculty position at a local college. Sam's job talk went well. The faculty was impressed by Sam's publications and teaching evaluations. As Sam was leaving, the former chair said, "You gave a really good interview. Now the faculty members have to decide if they want someone in your field or in another field." Sam was perplexed. When the faculty host for the interview contacted Sam, they said the faculty had decided to recruit someone in a different field. Sam was disappointed, but continued as a postdoctoral fellow and as a job applicant. Scanning the ads of the professional newsletter, Sam saw the same local college advertising for a person that looked like his CV. Sam contacted the person indicated in the ad and asked to reactivate his application. He was told to send in an updated CV and set up an appointment for an interview. Sam got the job and never knew the reason for the flip-flop in fields. Sam realized that each member of the faculty had their own busy professional careers in addition to the recruitment.

Be Yourself

When you go for an interview, you should be yourself. Perhaps you could fool a potential employer for a few days, but not after you accept the position. The worst thing you can do is to get the job for the wrong reason, not satisfy their needs, and leave both you and your new employer unhappy. You are being interviewed the whole time you are there.

Prepare for Your Visit

Before you visit, explore the website of the division, department, and institution/company/agency. This will inform your questions during your visit. You should carry out a literature search on those who are on your agenda. This will enable you to ask appropriate questions. Ask your colleagues and mentors about the people and the place. Make sure you know the kind of talk you are to give and who your audience will be. If someone at that institution has done work related to yours or has collaborated with you, be sure to mention this in your presentation. If the audience is small, don't voice your disappointment. Give a great talk and be as enthusiastic as if you were talking to thousands. Your hosts and the search committee will be there, and they will tell the others what a fantastic talk they missed. Be sure to leave time for questions. Your hosts will want to see how you handle questions, and you should be interested in the questions they have for you.

Be Honest with Yourself and Your Interviewers

Sam could not decide whether she wanted a job in industry or in academia. She applied for academic jobs as if that were her only interest, and she approached industry jobs in the same way. On a job interview at a university, the department chair asked Sam how committed she was to an academic career. Sam said she was totally committed to academics. The department chair asked Sam why she had interviewed with an industry representative at a recent national meeting. Sam was shocked and did not know how to answer. The chair had a friend working at one of the companies with which Sam had interviewed. When they were dis-

cussing their recruits, each mentioned Sam and then realized they were talking about the same person.

Be Nice and Know What You Want

You should be well rested before you go, because the visit will be rigorous with long days. Even if your hosts are not nice, you should be. Prepare questions ahead of time for each person you are meeting. People like to talk about themselves and their own work. You need to learn about them, the division, the department, and the institution/company/agency. However, don't look at your questions unless you are negotiating with the chair or division chief. Review the questions the evening before or the morning of your meeting with them.

Know what you want from this position in terms of salary and professional support. Your hosts may ask you on the first visit, but take your cues from them. If they don't bring it up, wait until the second visit.

You should be diplomatic. Some interviewers will attempt to get you to take sides in disputes of which you are not aware. Be careful and do not get involved in internal politics.

Be Diplomatic during Your Visit

Chris was applying for his first job as an assistant professor. While he was meeting with Dr. Jones, a senior professor in the department, she explained some of the problems she had been having with the department chair, Dr. Smith. A number of issues she raised were important to Chris. She maintained that Dr. Smith kept all of the departmental administrative resources for herself, that Dr. Smith discouraged faculty from traveling to professional meetings, and that Dr. Smith frustrated faculty attempting to pursue their own independent research interests. When

Chris met with Dr. Smith, Chris was very confrontational and accusatory. Unfortunately, Chris did not know that he had been set up by Dr. Jones, who preferred another candidate for the position. Chris played right into Dr. Jones's hand. Dr. Smith did not enjoy the meeting with Chris and had no enthusiasm for hiring him.

A Discussion Is a Shared Interaction—
Do Not Monopolize the Conversation

Remember, in our culture, a discussion generally represents a roughly even split in the overall time occupied by the verbalizations of each participant. Beware of monopolizing the conversation, especially if you are someone who tends to talk excessively when you are nervous. You should show interest in your potential colleagues and the setting. This is why knowing your interviewers' interests and having questions for them will be beneficial to you.

Remember Who Is Paying for Your Visit

When you are on a job interview, you are at your hosts' disposal. In general, you should not arrange to visit friends or family in the area. If you do, your hosts may feel that you have an ulterior motive and are taking advantage of them for free airfare. If you move to the area, you will have plenty of time to visit. On the other hand, some hosts may know you have family or close friends in the area. They may see this as an attraction for you and encourage you to visit them.

You should exercise moderation in all areas. In your interviews, be a good listener. With food be cautious: a big lunch leads to a sluggish afternoon, and a big dinner leads to a big bill. Take your cues from your hosts and do not order the most expensive meal. With drink the reasons for moderation are obvious.

Do Not Take Advantage of a Prospective Employer

Dr. Smith invited a prospective fellow candidate, Lynn, to travel to meet Dr. Smith's group and to give a seminar. The day before Lynn was scheduled to arrive, Dr. Smith called Lynn to clarify the schedule. Dr. Smith was told that Lynn had left the day before. When Lynn arrived, Dr. Smith inquired about Lynn's travel plans. Lynn had arrived two days early to be interviewed for another position. Dr. Smith suggested that Lynn split the cost of the airfare between the two potential job sites. Lynn was no longer in contention for the position with Dr. Smith.

Send Requested Materials Promptly

When you return home, you should send a thank-you note. If partners were involved in entertaining you, remember to include their names as well. You should promptly send any requested material, such as your plan or prospective budget.

Your Partner Is Part of the Interview Process

Your partner usually will be asked to accompany you on the second visit for a job. Your partner is also being interviewed throughout the visit. You should prepare your partner before your joint visit by discussing personalities, politics, and so forth.

If Your Partner Is Adamantly Opposed to the Move, There Is Little Purpose to the Visit

Chris was looking at her first real faculty position, an assistant professorship at a major university. Her first visit was very successful. She was invited to return for a second visit with her partner. Chris's partner, Lee, did not want to move. Lee especially did not want to move to the city where the university was located. Lee went on the visit determined not to have a good time. Although their hosts were very gracious, Lee chal-

lenged every positive statement made by the hosts about the university and the city. Lee found fault with every house they looked at, with every restaurant they ate in, and with every person they met. While the department was still favorably impressed with Chris, the faculty discussed whether they should invest any more resources or energy in her, since her partner was so negative about the move. They decided not to pursue Chris's recruitment. Further professional interactions were somewhat cold because the faculty felt that she had misled them.

NEGOTIATING A CONTRACT

With any position you take, you should get your offer in writing from your supervisor. It would be ideal if your supervisor's supervisor would also sign off on your offer. With two signatures, if one of the individuals should leave, retire, or step down from their current role, you still have a second individual in place to come through with what was promised.

If it is not in writing, it never existed and it won't happen. No matter how cordial the recruitment was, if the offer is not in writing, it was never discussed. When you receive a written offer, this is an initial offer. You need to be sure that all of your issues are addressed and included in the letter. Dollar amounts and dates are important and need to be specified. There are no do-overs here. You need to ask about issues that are important to you, but be sensitive to the financial realities of the person making the offer to you. If you take the job, you will have an extended, perhaps even career-long relationship with this person. You don't want to make this person angry before you agree to the terms of your position.

Do Not Continue to Visit When You Know You Will Not Take the Job

Lee was a hot commodity. Lee's subspecialty did not train many individuals. Those with Lee's credentials were heavily pursued. Company X really needed Lee's expertise to develop its new product line. Lee indicated an interest in relocating to Company X's area to be close to family. Lee was also enthusiastic about the promotion and the raise Company X was offering. Lee had two issues that needed to be resolved. Lee's partner had a good job and would need help getting a new job near Company X. Since Company X was in a totally different field from Lee's partner, there was little Company X could do about finding the partner a position. Lee's second issue was the depressed housing market for Lee's current home. Lee wanted Company X to buy the home. Company X was not in the real estate business and had never done this before. Lee maintained a strong interest in moving to Company X. Lee's family made multiple trips at Company X's expense to consider schools and homes in Company X's neighborhood. When Company X asked Lee to make a commitment, Lee responded that Company X needed to provide a position for Lee's partner and to buy their current home. When Company X responded that neither of those was a possibility, Lee withdrew from consideration. In retrospect, this long, drawn-out negotiation was never going to succeed. Lee wanted more than was possible for Company X. Company X had invested a great deal of time and money in this recruitment and had to start at the beginning again.

Get It in Writing and Be Prepared for Buyer's Remorse

You should get the offer in writing before you accept. Many colleagues have told us of verbal offers made and later withdrawn. In some cases, the verbal offer led to the candidate making significant life decisions, such as giving notice to their current employer, placing their house on the market, or buying a home in their future location. The conditions of the verbal offer may be understood differently by the two parties.

By having the offer in writing, you can examine the specific details and negotiate for clarity in those details. But recognize that it is impossible to have every detail in writing about every possible scenario that you may face in this new position. If you are too picky, you may begin to erode the trust your new boss has in you.

As you review the offer letter, remember that no job is perfect. Buyer's remorse (concern that your commitment is a mistake, analogous to the near-universal experience when one commits to buying a new home) is one of the standard, though transient, consequences of considering or accepting any offer. You should look for the situation that will provide the best environment for your professional growth and development, and a boss who is the kind of mentor who will have your best interests at heart.

The Disappearing Offer

Kim was excited by the offer of a promotion with significant funding to develop a new division. Kim's partner was very supportive of the move to this new institution in a new city. Kim did not have an offer in writing, but was told it would be arriving shortly. Each time Kim spoke to his future department chair, the offer got smaller and smaller. Eventually Kim learned that the offer was contingent on passage of a bill in the state legislature that would provide markedly increased funding for the institution. When that bill stalled in committee, the offer was doomed. Kim turned down the markedly lower written offer. Over the next two years, most of the faculty at the institution left for other positions.

SMOOTH TRANSITIONS—LEAVING

You should provide reasonable notice to your mentor, division chief, and chair. Some institutions require this in writing. This enables them to begin to search for a replacement and to organize the unit to assume your responsibilities in the interim. You should try to complete your responsibilities before you leave. For example, you should finish as many manuscripts as you can before you leave. It is easier to submit and revise manuscripts when you are at the same institution/company/agency as your colleagues.

You should leave on a positive note. You may want to return some-day. You should work to maintain the relationships you have established. It seems a natural tendency for some individuals to want to find fault with the institution they are leaving. They do not seem to understand that by taking such a negative attitude, they not only are burning their bridges, alienating former colleagues, and making it difficult to return in the future but are also running down an institution that positioned them for this move and will always appear on their CV. Colleagues at the former institution may be contacted for references in the future.

If someone is upset by your leaving, you should try to help them understand. Remember it is always easier to leave than to be left behind.

SMOOTH TRANSITIONS—ARRIVING

You should never speak ill of your previous institution—you used to be there. You should be positive about the adjustments you will make in your new setting. You should reach out to make new friends. You should work with your mentors and your network to learn as much as possible about your new setting.

Recognize that processes and rules will be different in your new institution. Some of these will seem inefficient and ineffective. You will become quite tiresome to your new colleagues if you constantly complain that your previous institution was better at this or that. This does not mean that you should not work to make changes, but if you attempt to change a process or rule, develop a positive rationale for this change. Also, attempt to find out who is invested in the status quo and assess the potential consequences of making the change.

One lesson we learned after our first move is, when asked what is our favorite place, it is always where we are now. Perhaps you moved there primarily for the professional environment, and some aspects of the locale (weather, city size, etc.) are not what you would have chosen. But others live there quite happily and of their own choice. Find out what the local appeal and opportunities are, and enjoy exploring your new home.

CHAPTER 10

Developing Time Management Skills and Short-Term and Long-Term Goals

FOCUS TO ACHIEVE SUCCESS

Different aspects of your career may progress at different rates. You need to take a realistic, overall view of your career and recognize that no one is an immediate expert in all aspects of academic life. Development of expertise in a very specific area may require dedication, one or more mentors, and time. Even with appropriate focus and productivity, recognition as an authority in a field requires time that is measured in years, at least five to ten usually.

DO NOT EXPECT TOO MUCH FROM YOURSELF TOO SOON

Professional competence in areas such as consulting or professional service is rapid, and the rewards are tangible. The danger is that your attention might be distracted by activities that preclude the development of other aspects of your academic career. Teaching and research require dedication and time to develop. Recognition for your research and publications may be delayed, and you must recognize that the gratification will be primarily personal. Rewards for teaching and career development may be even more nebulous. Committee work and administration can be very time-consuming and may go unrecognized. Each of these activities, however, is important to the development of your academic career and to the maintenance of the infrastructure of your institution/company/agency and your profession.

HOW DO YOU DEFINE SUCCESS?

Publications and Impact Factor

There are many different ways to be successful. Typically, the number of peer-reviewed research papers published with you as first or last author was considered important for demonstrating a successful research program. The message was that department chairs and deans could not read, but they could count—a reference to the difficulty of judging the importance of publications. However, there has been an attempt to judge the significance of a paper by the impact factor of the journal in which the paper was published. A journal's impact factor is related to the average number of times each paper published in the journal is cited in other publications. However, the most frequent number of citations of each paper in a journal is 1. Higher-impact journals will have a few papers that are cited very often, which raises the average number of citations for the journal as a whole.

Another parameter for ranking journals is the Eigenfactor (see www.eigenfactor.org), which ranks the importance of the journal to scientific knowledge formation using a ranking algorithm similar to the way a search engine (e.g., Google) ranks websites. The Eigenfactor uses a broader network approach to ranking a journal's importance; for example, if a natural science journal impacts the natural sciences through its network of influence, this effect is included. The Eigenfactor recognizes citation differences between disciplines and adjusts for these differences, thereby permitting more equitable comparisons between disciplines. The results are given as Eigenfactor and Author Influence scores, and each of these is also provided as a percentile ranking.

You can also calculate your own impact on the field by how many times your papers have been cited. This is perhaps a more appropriate determination of the significance of your work. Another statistic for evaluating the influence of your publications is the h-index, which is a different way of looking at your number of publications and the number of times they are cited. You can calculate your h-index using tools at Scopus, Google Scholar, and other sites.

Aiming High, Falling Short

Kim was thrilled to be a postdoctoral fellow in Dr. Jones's group. Dr. Jones would publish only in Cell, Nature, or Science. Kim planned on a two-year postdoc before moving on to a tenure-track academic position. After six months, Kim realized that having enough data for a first-author Cell paper would require seven years of research. If Kim left before completing seven years, Kim would only have a middle authorship. Kim decided to move to Dr. Smith's group, with which it was possible to have a first-author paper in a reputable but less prestigious journal after two years.

As a mentor, you should recognize the importance of considering the publications, grant funding, and leadership achievements of your trainees and their impact on the field.

Grants and Patents

Another definition of success is the number of grant dollars you secure. You could also consider the number and types of grants. Some research grants, such as the NIH R01, are considered more prestigious than others. Program projects or center grants that focus on a theme, include multiple investigators, and provide support for core resources are also important. Institutions especially appreciate training grants that provide salaries and some research support for graduate students, postdoctoral fellows, and/or junior faculty. Training grants reduce the drain on mentors' research grants and therefore extend their research budgets. Participation in obtaining training grants, grant writing and securing funding are all important for mentees.

Success can also be measured in terms of patents and start-up companies. In addition to grants, institutions and companies are looking to intellectual property in the form of patents to diversify their research funding portfolio. When appropriate, mentees should be included in patent applications.

Speaking Invitations and Other Honors

The successful investigator is invited to speak at professional meetings and at institutions around the nation and the world. Some of these talks are named lectures that have been endowed in honor of a particular individual. Other presentations are part of honorific awards from a professional organization or foundation. If you are invited to another institution to give a talk (e.g., a seminar or grand rounds) and give an excellent presentation on a topic in which you have developed expertise, this will add to your credibility and should be recorded in your CV.

Another dimension of success is election to honorific societies. This may involve nomination by colleagues who are already members and a vote by the entire membership. Mentors should nominate mentees for every appropriate award and society membership.

Nominate Trainees for Awards Early and Often

Kelly was a postdoctoral fellow in Dr. Taylor's group. They were preparing an abstract for the national professional meeting. Kelly asked if Dr. Taylor would nominate her for a special award. They reviewed the qual-

ifications for the award and decided to wait until the next year to have a stronger application. At the professional meeting, they attended the talk of the award winner. What a disappointment! Kelly was much more qualified than the awardee. From this point on, Dr. Taylor nominated every trainee for every possible award. The next year, Dr. Taylor nominated Kelly, and Kelly was selected for this honor.

Moving Up the Ladder in Institutions, Companies, Agencies, or Professional Organizations

For those who enjoy supporting the academic infrastructure, success can also be measured by participation on committees and in leadership roles. You should pursue only those opportunities that are important to you, in which you feel you can make a difference, and that will fully engage you. Leadership is articulating a vision and setting an example for others. Promotion, for example, to division chief or department chair leads to increasing responsibilities, as does chairing a committee or election to an office in a professional organization. Along with increasing responsibilities comes the opportunity to mentor a larger group of individuals.

The Success of Your Mentees Is a Measure of Your Success

The most enduring measure of your success is the continuing achievements of your mentees. Their publications, grants, patents, honors, promotions, and offices are testaments to your successful mentoring. The true mentor cherishes the mentees' successes and is not envious of or threatened by them.

Your Success Is Measured by the Success of Your Trainees

A professional organization provides an award to the first author of the most significant paper in the organization's journal for the previous calendar year. The award includes travel to the annual meeting, a small prize, and the opportunity to give a talk. One year, the senior author of the selected paper was upset that he was not the winner and wanted the rules changed. He remained more interested in adding to his own awards than in celebrating the success of his trainee.

Set Goals and Evaluate Them Annually

Realization that you are making progress in different areas of your career may require you to look back one year or even five years. To ensure development of the various aspects of your career, you should set both short-term and long-term goals. A long-term goal, such as being promoted to associate professor and receiving tenure, is vague and nonspecific. You stand a better chance of achieving these goals, for example, by determining the requirements for promotion and tenure in your institution, and planning how to meet these expectations by meeting short-term goals.

You should look backward as well as forward. You need to see how far you have come. When you are busy with the details of your life and career, it is hard to recognize your achievements unless you take the time to reflect upon how well you are meeting your short- and long-term goals.

Review your goals annually. You may be surprised at how much progress you are making toward your longer-term goals. If you find that progress is lagging in one or more areas, careful evaluation and review with your mentors will help you develop better strategies for success.

You should not let temporary setbacks frustrate you. Achieving goals always seems to take longer than you plan. Given this reality, build some extra time into your plan.

CHAPTER 11

Developing Your Grant-Writing Skills

SELECTING GRANT OPPORTUNITIES

Funding for your research, whether it is in the area of basic science, translational, clinical, or clinical trials research, is essential for you to achieve your goals. In this chapter, we address the NIH style of grant writing. You should know that the same fundamental principles hold true for all types of proposals, whether they are for foundation grants or clinical trials proposals. The formats may differ; for example, some foundations may require only a letter. Talk to a representative of the funding source to determine if your idea has any viability and go to the source's website to obtain details regarding format and submission.

Cultivate Your Network of Program Officers

When you begin the search for grant support, your best allies are the agency representatives or program officers. They can inform you of upcoming requests for proposals (RFPs) or requests for applications

(RFAs). They can tell you which type of application is most relevant to your area and to the stage of your career. They can also tell you what type of feedback you will receive. Should your proposal not succeed, program officers can help you understand the critique of your proposal and why you did not receive support. With their insight, you can restructure your proposal and submit a responsive revised proposal with an improved chance of success. Private foundations may simply inform you of your success in securing funding.

Be Sure There Is Funding for a New Initiative

Contact the agency representative to determine if the agency is interested in receiving your proposal and if it has money for a new RFP or RFA. A positive response does not guarantee that you will be funded. You can also ask how many proposals will be funded in the current competition.

There have been new initiatives that were announced, but no proposals were funded because there was no money for the program. There was never any intention of funding. The announcement was a trial balloon to demonstrate interest in the topic to key individuals in the agency. Your time is too precious to waste it on an application with no chance of funding.

Meet Short Deadlines—Do Not Procrastinate

If your agency contact agrees that you should apply to a new program and there is money, you should apply even if there is not a lot of time. The first time an announcement is made, there may be less competition due to the inability of some to mobilize and submit a proposal on time.

Your contact may also make you aware of an upcoming announcement before it appears. This will give you a running start on your proposal and an improved chance for success.

How Do You Find Out about Program Announcements?

To determine which agencies would consider funding your work, consult the following sources: your mentors, your colleagues, the acknowledgement sections of published papers, conflict-of-interest declarations for talks and posters at professional meetings and seminars, your grants and contracts office, and the Internet. You should visit the websites of agencies identified by any of the sources noted above to learn more about the agency, how to contact its representatives, current funding priorities, types of support, blank proposal forms, currently funded proposals, and program announcements.

The following websites may be helpful:

National Institutes of Health (NIH) Guide for Grants and Contracts, http://grants1.nih.gov/grants/guide/index.html

NIH Unsolicited or Investigator-Initiated Applications, http://grants1.nih.gov/grants/guide/parent_announcements.htm

NIH Research Training and Research Career Opportunities, http://grants1.nih.gov/training/index.htm

NIH Small Business Research Funding Opportunities. http://grants1.nih.gov/grants/funding/sbir.htm

Health Resources and Service Administration, http://www.hrsa.gov/grants/

Agency for Health Care Research and Quality, http://www.ahcpr.gov/fund/grantix.htm

National Science Foundation, http://www.nsf.gov/funding/

Robert Wood Johnson Foundation, http://www.rwjf.org/applications/

March of Dimes and Birth Defects Foundation, http://www.marchofdimes.com/professionals/691.asp

Centers for Disease Control and Prevention, http://www.cdc.gov/about/business/funding.htm

ORGANIZATION OF THE NIH

Develop Familiarity with the NIH

We are most familiar with the NIH since much of the support for research in departments of pediatrics comes from the NIH. The NIH consists of two arms, Program and Review. These were designed to be distinct and separate in order to disconnect decisions about the programmatic needs of the NIH from decisions about the scientific merit of, and funding for, individual applications. Program includes the various Centers and Institutes (CIs) that determine funding priorities and issue requests for proposals. Review includes the Center for Scientific Review (CSR) and the review groups. There are certain proposals that are reviewed in CIs, but this is rare. Helpful websites include those of the NIH CIs (http://grants.nih.gov/training/trainingfunds.htm) and the NIH Center for Scientific Review (http://grants.nih.gov/grants/peer/peer.htm).

You should establish relationships with the relevant program officers in the appropriate CIs at the NIH—that is, those with priorities that include your area of research. They can advise you regarding the relevance of particular program announcements or certain types of proposals. Each CI will have specific RFPs/RFAs, unique funding mechanisms, and particular interpretations of training and career development awards.

CI program officers will be able to assist you in selecting the relevant review group for your proposal. Your mentor(s) and senior colleagues in your research area can also provide insight into review group selection. If you are responding to an RFP or RFA, the review group will already be determined. In this case, review may be carried out within the CI.

NIH Program Officers Are Valuable Resources

Kelly was an MD, PhD postdoctoral fellow with plans for a career in academic medicine. His mentor suggested that he apply for a K08, a five-

year grant that could begin during fellowship training and transition into the early years on faculty. His mentor's research was heavily supported by Institute A. Kelly planned to apply to Institute A for his K08. He repeatedly tried to contact the program officer for Institute A K08s, but was not successful. His mentor attended an Institute A program and learned that Institute A, at that time, did not award K08s to MD, PhDs. His mentor immediately contacted Kelly and urged him to contact the K08 program officer at Institute B. Kelly was able to reach the Institute B program officer and learned that Institute B did accept K08 applications from MD, PhDs. He was awarded a K08 by Institute B. Kelly was fortunate to learn about Institute A's K08 policy before he submitted his K08 to the wrong CI. Kelly has become a tenured associate professor, and his program officer stays in touch, for example, by visiting him at his poster when he presents at national meetings.

Choose the Best, Not the Easiest, Review Group for Your Proposal

To select a review group for your proposal, you should consider the descriptions of review groups that seem appropriate. Does the description mesh with your research plan? Are you familiar with the research of the members of the review group? When you review the publications of the members of the review group, do you find several individuals who publish in similar areas and who ask questions similar to yours? If they are interested in a similar area of research, this may be a good choice. Your proposal should cite their work if it is appropriate.

Don't be frightened by tough review groups. Experience has shown that they typically provide fair, constructive reviews. "Easy" review groups may be more arbitrary, less consistent, and less informative. If you find a review group that is a good match with your research, you should indicate your choice in the cover letter you submit with your proposal. Present the rationale for your choice using keywords found in the description of the role of each review group.

You are in a far better position to understand your research program than is someone performing a cursory review and triage at the Center for Scientific Review.

Constructive Criticism from a Tough Review Group Will Lead to Eventual Funding (If You Pay Attention to the Comments)

Lee had a funded NIH research grant. Review Group A reviewed Lee's proposal and the amended proposal was scored in the fundable range. Review Group A had a reputation for being tough but fair. Lee was preparing an application on a slightly different topic but with methodology similar to that of his funded grant. Lee was planning to request Review Group A, but a senior colleague, Dr. Smith, suggested that Review Group B was easier. Dr. Smith received a score in the fundable range from Review Group B on the initial application and was funded. The members of Review Group B were all long-time colleagues of Dr. Smith. Since Dr. Smith was one of Lee's collaborators on the new proposal, Dr. Smith felt that this would improve Lee's standing with Review Group B. Lee decided to request Review Group B. Unfortunately, members of Review Group B did not consider Lee to be the person to perform the proposed research. They did not provide any constructive criticism of Lee's proposal and did not score it. This placed Lee's proposal outside of the fundable range. Lee was stunned by the capriciousness of Review Group B. Without any guidance from Review Group B, Lee did not know how to amend the proposal to receive an improved score. Lee requested Review Group A for his amended proposal. While Review Group A did not give Lee a fundable priority score, it did provide excellent feedback. Lee decided that all future applications in this research area would go to Review Group A.

You May Request Exclusion of a Reviewer

What do you do if the review group you've selected includes a bitter professional rival? When you receive notification of your review group assignment, e-mail the scientific research administrator (SRA) for the review group to request that that individual not be allowed to review your proposal and not be in the room when the review group discusses it. This can be done even if the person is the chair of the review group. Do not be so naïve as to assume that the person would rise above your past disputes and give an objective evaluation, or that they would openly declare the conflict and disqualify themselves from the review. This may be just the opportunity they have been looking for to discredit your research program.

Do Not Trust Your Nemeses to Recuse Themselves

Chris and Kim were colleagues in the same department. Although their topics of research were different, they did use similar methods and chose the same review group for review of their grant proposals. By chance, they submitted research grants to this review group at the same time. When looking over the list of review group members, Chris noted that Stacey was a member. Both Chris and Kim had previously collaborated with Stacey. Both of them ended their collaboration with Stacey after he took major credit for their joint projects. Chris decided that when the appropriate review group was assigned to her grant proposal, she would ask that Stacey not be allowed to review her application and that Stacey be asked to leave the room during the discussion of her proposal. Kim laughed and told Chris that she was overreacting and that Stacey would declare the previous collaboration to be a conflict of interest and would disqualify himself as a reviewer. Kim felt that even Stacey would not be so unprofessional as to purposely give a bad review to a former collaborator. Both Chris and Kim were assigned to Stacey's review group. Chris wrote asking that Stacey have nothing to do with the review of her grant, but Kim did not. Chris received a reasonable review with a priority score in the fundable range. Kim's review was very negative, very personal, and not at all constructive, and her score was far outside of the funding range. Chris received support for her proposal. Kim had to revise and resubmit.

When Kim was assigned to Stacey's review group for the amended appli-
cation, Kim requested that Stacey not be allowed to review her grant and
that Stacey be out of the room when her grant was discussed. This time,
Kim received a positive review, a priority score in the fundable range,
and funding. While Kim will never know who reviewed her initial sub-
mission, she did know that Stacey did not review her second version. Kim
also noted that Chris was funded months before she was. She resolved that
she would immediately request that Stacey not be allowed to review any
future proposals.

To Improve Your Funding Chances, Request up to Three CIs

In addition to specifying a review group in the letter of transmittal
with your proposal, you should also specify up to three CIs that would
be interested in funding your proposal. While the program officer
you have been talking to represents only one CI, they may be able
to suggest other CIs with similar funding priorities. Your mentor
and senior colleagues will also have helpful suggestions. The publications and presentations of others in your research area should specify the CI supporting their research, and you should use their
acknowledgements to guide you.

CIs vary in the amount of funds they have and in their ability to
fund specific areas. Funding priorities of the CIs are listed at http://
www.nih.gov/icd. If you name more than one CI and your first choice
does not have the money to support your proposal, another CI may
have more resources and be able to step in and provide funding. The
program officer will help you complete the necessary paperwork for
multiple CI assignment.

Multiple Institute Assignments Pay Off

Sam was submitting a research grant to the NIH. Sam knew that Institute
A provided most of the support in his department. He indicated Institute

A as his first choice. However, Institute A was relatively impoverished compared to Institute B, which also covered the topic area of Sam's application. Sam requested both Institute A and Institute B in his cover letter. Sam received a reasonable funding percentile on his proposal. Institute A's funding line was 2 percentile points lower. Institute B funded to a level 3 points higher than Sam's score. Sam's proposal was funded by Institute B. Sam would not have been funded if he had not included a request for both Institute A and Institute B in his cover letter.

Submit a Courtesy Copy to Your Program Officer

You should e-mail a copy of your proposal to the program officer who helped you with your application. Include a statement indicating the review group and CIs you specified in your cover letter. Keeping the program officer informed is important, but it also serves as a check on the NIH grant triage system. You should receive notification of assignment to a review group and CI within four to six weeks after the proposal due date. You can also check the progress of your proposal on eRA Commons (http://era.nih.gov/commons/index.cfm). If you do not receive notification, you should contact you program officer and ask them to check on your proposal. If you are assigned to a different review group or CI than you requested, you should immediately contact the Center for Scientific Review and state your case.

Good Manners Pay Off

Lee was submitting her first grant proposal. Her program officer had been very helpful and seemed genuinely interested in her project. Lee e-mailed a courtesy copy to her program officer. Somehow her electronic submission of her proposal was not successful. The program officer contacted Lee as soon as the program officer realized that there was a problem with Lee's submission. Since the program officer had received the proposal before the deadline, Lee's proposal was considered in that round. She did not have to wait until the next competition to submit her proposal.

Long-Term Goal: Establishing a Research Career

Table 4 shows how to begin, as a first-year postdoctoral fellow, to establish a research career. While this is meant as a general guideline, note that at all times you need to have time for research and publishing the results of your research.

WRITING A GRANT: SELECTING THE SPECIFIC AIMS, PREPARING THE BUDGET, AND DEVELOPING THE RESEARCH PROPOSAL

You Will Never Get a Grant for Which You Do Not Apply

We enjoy some success in obtaining grant support for research and training. However, our success rate is probably no higher than most. We follow the same instructions as those coined for voting in a certain city: we just apply "early and often." Once when we were celebrating the funding of a proposal, a colleague suggested that we must be getting every grant for which we applied. We invited him to see our file drawers full of rejected grant applications. There is no guarantee that you will be awarded every grant for which you apply. However, one thing is certain: you will never receive a grant if you do not apply for it.

Success Requires Confidence, Focus, and Persistence

Important factors for success in obtaining grants are confidence, focus, and persistence. You need to be confident enough in your ideas and methodology to submit a proposal and, if you are initially unsuccessful, to respond to the critique and submit a revised pro-

Table 4. *Timeline for Establishing a Research Career*

	Fel 1	Fel 2	Fel 3	Fac 1	Fac 2	Fac 3	Fac 4	Fac 5	Fac 6	Fac 7
Select mentor	✓			✓						
Apply for individual training grants	✓	✓								
Do research	✓	✓	✓	✓	✓	✓	✓	✓	✓	✓
Publish	✓	✓	✓	✓	✓	✓	✓	✓	✓	✓
Apply for grants to transition from fellow to faculty		✓	✓							
Apply for faculty positions			✓							
Establish affiliations with centers, CIs, and training programs				✓	✓	✓	✓	✓	✓	✓
Apply for funding as part of a program project				✓	✓	✓	✓	✓	✓	✓
Apply for funding as PI of a program project, center, or training program									✓	✓

NOTE: Numbers are years. Abbreviations: CI = Center or Institute; Fel = fellow; Fac = faculty; PI = principal investigator.

posal. Your proposal should focus on a single area, one in which you have the experience, preliminary data, and publications necessary to support an application. You need to demonstrate your persistence by revising and resubmitting your proposal for the next cycle, with new data and publications.

Do Not Let a Bad Review Shatter Your Confidence

Stacey was very successful throughout school and training. Because he enjoyed the research required as part of his training and found teaching very rewarding, Stacey decided to pursue an academic career. During his training program, Stacey, with the help of his mentor, submitted a grant proposal to support his research. Stacey's previous proposals for training grants and travel grants for funding to attend professional meetings had all been successful. When the review came back, Stacey was devastated. The review began with a scathing condemnation of his mentor and proceeded to destroy point by point Stacey's preliminary results and research plan. In addition to the absence of any constructive criticism, there were a number of glaring errors of fact in the review. Stacey's mentor suggested that the review was a personal attack on her, not Stacey, his preliminary data, or his proposed research. She argued that since there were many factual errors, Stacey's request for a re-review would be honored by the NIH. Despite her support, Stacey did not request a re-review and left research. Stacey allowed a vindictive review committee member to rob him of the joy he felt in doing research.

Preparing a Grant Requires an Organized Plan

The following questions must be addressed as you plan and prepare any grant proposal:

- · Who?
- · What?
- · When?

Considering these issues formally and effectively early in the process will clarify your plans for your application.

Who?

This question is not trivial. The principal investigator (PI) is the person responsible for writing the proposal and carrying out the research. For experience, the trainee should write their own proposals for training support. It is generally quite easy for reviewers to distinguish an application prepared by a trainee from one written by a more seasoned investigator, so the trainee should not submit a copy of the PI's proposal. Part of the training experience is preparing grant applications so that the trainee will be able to secure independent funding before the end of the training period.

We have all seen instances in which the person writing a grant was not listed as the PI. For purposes of promotion and tenure, a junior faculty member needs to have independent grant funding and an independent research program. The junior faculty member should be the PI for most of the grants he or she writes. Early in your research career, as a fellow or very junior faculty member, you may prepare a grant providing salary for yourself that will be submitted listing your mentor as the PI. This is acceptable as long as there is a very specific plan to develop an independent research career for the fellow or faculty member, and the mentor gives credit for the junior member's contribution in letters of recommendation.

Acknowledging the need for collaborators to perform today's complex research, the NIH has developed the concept of co-equal co-PIs. Co-PI status indicates an equality of expertise, experience, and publications. Each of these individuals could submit their own independent research grant, but they have chosen to collaborate on this particular project. Hopefully promotion and tenure committees will recognize each co-PI as a PI and reward them appropriately with promotion, tenure, and space.

In addition to establishing the PI or co-PIs, potential collaborators need to be identified. These are individuals who have specific skills in methodology or who direct core facilities providing services needed for the proposal. You cannot be an expert with demonstrated published skill in every method. You will need to rely on collaborators to teach you certain methods or to provide certain services in order to accomplish your proposed research. You need a biographical sketch and a letter agreeing to be a collaborator from each person you identify for this role.

Modern Science Requires Collaborators

Sandy was applying for her first R01. Since the R01 is an independent research grant, Sandy mistakenly concluded that she needed to do all the work herself. She prepared a very careful and thorough proposal based on her previous work in the area. The proposal involved a variety of methods, many of which Sandy had not yet attempted. She was confident that she could master these new methods. However, they required a great deal of equipment that Sandy did not have. She included requests for the equipment in her proposal. The equipment budget was excessive. Sandy barely finished the application in time to meet the submission deadline. She had not asked her mentor, Dr. Smith, or other senior faculty members to review her proposal before submitting it. After the proposal had been submitted, Sandy gave a copy to Dr. Smith. Dr. Smith suggested to Sandy that "independence" did not mean working in a vacuum but, rather, referred to research design, especially creativity in crafting and testing hypotheses. Dr. Smith emphasized that Sandy's research would be seriously delayed by the investment of time required to order all of the equipment, set it up, and establish the methods in her lab. Dr. Smith informed Sandy that there was a core facility on campus that provided method A in a cost-effective fashion with high-quality output. Method B could be accomplished in the laboratory of a colleague at another institution who had established the method after ten years of development.

Method C had been developed by Dr. Jones in that department. Dr. Jones could teach Sandy this method, and Sandy could use Dr. Jones's equipment. Dr. Smith explained that Sandy should have listed these scientists as collaborators, obtained biosketches and letters of collaboration from each of them, and included these in the proposal. Sandy's grant was not scored, and the review group echoed Dr. Smith's recommendations. By following these suggestions, Sandy obtained a fundable score for the next submission, and her amended application was funded.

Key Personnel Must Be Specified

If on the grant you are requesting salary support for key personnel in your group in addition to yourself, you need to demonstrate their expertise through their NIH biographical sketch and in the budget justification section of the application. Providing a named individual in your group who has experience and publications indicates that you are already working in the area and will continue to do so with the funding from this grant.

What?

In answering this question, you first need to decide on your research topic and determine whether the topic is appropriate to the request for proposals and/or the mission of the funding agency to which you are applying. You then need to determine the specific aims of your proposal. It must be possible to accomplish these focused research objectives within the time frame of the grant funding period. The specific aims are determined by the hypotheses you want to test. Nearly all NIH grant proposals are hypothesis driven. The NIH has been much less interested in supporting technological breakthroughs except as they support hypothesis-driven research or particular mandates, such as the Human Genome Project, or certain NIH director's "Roadmap" initiatives, such as in epigenetics.

The specific aims detail what the proposed research is intended to accomplish. Each specific aim should target a key scientific question. A five-year proposal should include between three and five specific aims. You should use your specific aims to organize the rest of your proposal, since this will provide a constant framework to assist the reviewers in following the structure, logic, and flow of your application. Therefore, take time to develop and refine your specific aims before moving on to the other sections of your proposal.

When?

You must divide the goal of submitting your grant into shorter-term subgoals to be sure you meet the deadline. You should use table 5 as a guide to planning.

You may not always have five months to prepare a grant application, but your initial application will always take longer than you expect. Some investigators like to leave everything to the last minute. You cannot do this and expect to meet your deadline with a complete, thoughtful, well-organized, and neatly prepared application. Preparation of a grant proposal is a test of your ability to plan and organize in order to accomplish the research you are proposing and an opportunity to demonstrate these positive traits to the reviewers.

PREPARING THE GRANT PROPOSAL

The Specific Aims Represent the Organizational Structure of Your Grant

The grant proposal should be based on, and organized around, your specific aims. These aims need to be hypothesis driven and testable. They should be based on a new idea and should be brief and clear. They serve as the organizing outline for each of the other sections of your proposal. Remember, the person reviewing your grant is also reviewing many others. Anything you can do to improve the clar-

Table 5. *Five-Month Timeline for Preparing and Submitting a Grant Proposal*

	Month 1	Month 2	Month 3	Month 4	Month 5
Obtain application and read instructions	✓				
Discuss interest in proposal with program officer	✓				
Literature review	✓	✓	✓	✓	✓
Develop specific aims	✓	✓			
Discuss specific aims with mentor and colleagues	✓	✓	✓		
Perform preliminary experiments	✓	✓	✓	✓	✓
Write background and significance	✓	✓			
Write preliminary results		✓	✓	✓	✓
Write research plan			✓	✓	✓
Collect biographical sketches			✓		
Request letters of collaboration			✓		
Review proposal with mentor and colleagues				✓	
Make changes recommended by mentor and colleagues					✓
Draft letter of transmittal indicating type of grant, RFA or RFP, CIs, and review group					✓
Reread instructions and assemble application					✓
Send courtesy copy to program officer					✓
Contact SRA regarding due date for update before review					✓
Send update of progress on research program when appropriate					

ity of your proposal will aid the reviewer in following your thought process and understanding your planned research.

The Specific Aims Should Be Focused but Independent of Each Other

One of the major criticisms of junior investigators is that they propose a lifetime of work in a two-year proposal. The specific aims should be reasonably accomplished within the time frame of the grant proposal. Another major criticism is that the specific aims are structured such that if the first specific aim is not achieved, no work can be done on the other specific aims. The aims need to be interrelated and focused on achieving a common goal and testing a single overarching hypothesis, but they need to be independent, so that the second and subsequent specific aims can be carried out even if the first is not achieved.

Specific Aims That Are Sequentially Interdependent Will Be Found Fatally Flawed in Review

Morgan prepared a grant proposal with three specific aims. The first involved the development of a knockout mouse with a particular phenotype that would mimic a human disease. The second specific aim provided developmental characterization of the affected mice. The third aim attempted gene therapy for the knockout mouse. Upon reviewing the proposal, Morgan's mentor, Dr. Mason, noted the dependence of specific aims 2 and 3 on specific aim 1: if the mouse could not be made to have the desired phenotype, the aims could not be achieved. Morgan stubbornly refused to rewrite the proposal and submitted the one Dr. Mason reviewed. The review group did not score Morgan's proposal, citing the impossibility of accomplishing any of the proposed research if Morgan could not generate the knockout mouse with the human phenotype. They went on to list a number of human disorders for which there was no appropriate mouse model, in spite of multiple attempts by many different research groups.

Emphasize the Significance of, and the Rationale for, Your Proposal with a Balanced Review

The Background and Significance section is not simply a review of the literature but, more importantly, provides the rationale for your proposal. The rationale will be more compelling if it states the logical foundation for each specific aim. Rather than leaving it to chance whether the reviewer will recognize the argument for each specific aim, make it clear by organizing the Background and Significance section according to the specific aims. Your goal is to convince reviewers that your proposed research is important and that one outcome of your research will be to advance knowledge in the field. You want to clearly demonstrate the novelty and potential significance of your proposed work. As part of your literature review, determine whether any members of the review group you are requesting have published in the area of your research. Their relevant publications should be cited in your proposal. If you don't cite them, they will assume that you do not think their work is important. This could affect their review of your proposal. Similarly, provide a balanced review and cite the publications of others equitably, or the reviewers may be concerned about your ability to be objective in your research.

Read the Primary Literature and Go Back More Than Five Years

Taylor was in a hurry to write and submit a grant proposal by the deadline. Lee had time only to do an online search of the literature and focused on the articles published within the past five years. Rather than reading the papers herself, she read only the abstracts. Taylor did not ask anyone at her institution to review her proposal before submission. The grant reviewers noted the superficiality of Taylor's literature review and that she had cited only articles published in the past five years. They chastised her for proposing experiments that had already been performed and published. They pointed out that Taylor did not correctly interpret a number of the cited studies. Apparently, the abstracts had not provided Taylor with enough background to correctly evaluate the research. They were concerned that Taylor did not address some of the basic controversies in the area that had been presented in review articles. The reviewers did not score Taylor's application. Taylor was very embarrassed. She should have undertaken a more thorough review of the literature, had her mentor and senior colleagues review the proposal, and submitted the proposal for the following deadline. Submitting a proposal before it is ready does not save time. It usually puts you behind schedule. It also does not enhance your reputation. When these reviewers see Taylor's name in other contexts, such as reviewing her abstracts for presentation at scientific meetings or her manuscripts for journals, they will remember Taylor's very superficial literature review and wonder if she is superficial in the performance and evaluation of her own research.

Preliminary Results Demonstrate Ability and Feasibility

In the Preliminary Results section, you should demonstrate the feasibility of your specific aims through your own research. As with other sections, it is best to organize your Preliminary Results according to your specific aims. This section also showcases your relevant expertise, particularly with state-of-the-art techniques. You need to show that you are the person to perform this research and that you are already doing so. The data should be unequivocal, should be presented clearly, and should include a critical discussion of any of their limitations. It is important to have a substantial volume of preliminary, unpublished data showing your momentum in this research area. However, if you have so much data that you could develop one or more publications from the material in the Preliminary Results section, you may be criticized for not being able to generate publications from your work. Figures, graphs, and tables are helpful to support your scientific credibility. You should include color figures in the proposal if they are the best presentation of your data.

Preliminary Data Are Critical for Successful Review

Chris and Lynn are each preparing a grant proposal. Chris includes detailed descriptions of the methods in the Proposed Research section and doesn't have room for figures and tables summarizing his data from preliminary research. Lynn incorporates data from preliminary studies that illustrate her ability to perform these methods, so she just mentions the standard methods she has already demonstrated she can accomplish. The reviewers fault Chris for too much experimental detail and not enough preliminary results. Lynn is congratulated by the reviewers for her preliminary data. Lynn is funded; Chris is not.

The Research Design and Methodology Section Should Not Be a Cookbook, but Should Demonstrate Your Logical Approach

Your Research Design and Methodology section should be organized according to your specific aims and should be presented in enough detail to permit reviewers to evaluate what you will do and why. Each specific aim needs a rationale. You also need to discuss what you will do if your hypotheses are not confirmed. Do not detail standard methods that have already been published. It is best if you can demonstrate that you or your collaborators have published experiments using the methods and technologies proposed. Reviewers are always concerned whether the investigator will know how to process data and design future experiments. To be sure that you remember to discuss the logical progression of your data, you may wish to include separate sections entitled "Anticipated Results" and "Alternative Approaches" at the end of the presentation of each experimental design for each specific aim or subaim.

WRITING STYLE

Neatness Counts

In writing your proposal, you want to be very clear. One reason to have colleagues review your proposal is to help you find grammatical errors that may obscure your science. You should spell-check your proposal because misspellings will be considered a sign of sloppiness that may carry over from your writing to your research.

Try to Teach, Not Snow, Your Reviewers

Your proposal needs to be written in straightforward language, without jargon. You need to realize that while the reviewers may be in your general field of interest, their expertise is probably not in your specific area.

Avoid "Walls of Text"

The appearance of your proposal is important. A page full of text with no spacing is daunting. You should break up the text with spacing, headings, schematic diagrams, and, best of all, data (figures and tables).

BUDGET JUSTIFICATION

The Budget Justification Needs to Be Realistic

Most grants have either a specified dollar amount or range. You do not have to show how you will spend every dollar. However, you do need to justify personnel. The budget justification represents another opportunity to demonstrate that you are organized and knowledgeable and will be able to carry out the proposed research.

The Budget Justification for Personnel Needs to Assure Reviewers of Success

You should include a salary for yourself if it is allowed and indicate that you have a level of effort adequate for the project and that you will be able to devote that level of effort. Some grants specify the amount of effort you need to commit. Discuss this level with your department chair to ensure that you will be able to fulfill the requirements of the grant if your proposal is successful. The grant is a contract between your institution/agency/company and the granting agency. You must be able to fulfill your time commitment if the proposal is funded.

You do not want your percentage of effort at too low a level, or it will be assumed that you will not be able to devote sufficient time and energy for the success of this project. Your effort should be 20 percent or more for an NIH R01 or similar project.

Some senior investigators have so much grant support that they need to be careful that their grant funding does not exceed 100 percent effort. The NIH computer database makes it easy to identify individuals with more than 100 percent effort.

The other personnel you list should be appropriate to the project. Since you need extensive preliminary data to get a research grant, it is likely that you already have one or more individuals in your group who helped you acquire these data and who has the appropriate experience. Administrative personnel are usually not permitted. Consultant costs are generally discouraged unless you have a compelling justification for them and can communicate this well. It would be better to make the person a collaborator at no salary or percent effort, or include them as a co-PI if their contribution will be substantial.

You should explain the role of each person in detail. You should include their previous training and experience relevant to the project.

IF YOU HAVE TO PREPARE A DETAILED BUDGET

The Equipment Should Be Used Solely for This Project

Any equipment requested should be specific to the project. The use and importance of each piece of equipment needs to be justified. Reviewers maintain that the institution should provide laboratory setup equipment for new investigators. Reviewers generally expect that equipment will be purchased during the first year, so any exception needs to be carefully justified.

Supplies Should Be Discussed by Category

Supplies should be itemized by category—glassware, animals, chemicals, radioactivity, and so forth—and their use described. In general, you are allowed between $10,000 and $15,000 per full-time equivalent (FTE) in personnel. To determine how many FTEs you have, total the percent effort. If you have large animal or tissue culture costs, you can justify higher amounts.

Travel Should Be Limited and Focused

On NIH grants, only domestic travel for the principal investigator is permitted, at the rate of $1,000 to $1,500 per trip. Travel should be to a national meeting where you are presenting your results.

Other Expenses Should Be Carefully Considered

Other expenses can include publication costs (journal page charges, color figures), warranties for equipment, radioactive and toxic waste disposal, and service contracts. Typically telephone, fax, postage, and overnight delivery are not covered. Patient care, alterations, renovations, and consortium/contract costs are generally frowned upon. Before you include these costs, you should discuss them with the review group's SRA and your program officer. If they say that these expenses are acceptable, you must carefully and explicitly justify them.

Any Drastic Changes in Budget from Year to Year
Will Attract Reviewers' Attention and Concern

There should be no major equipment purchase after the first year, unless there is a very specific justification for it. The NIH does allow for a 4 percent increase each year in all categories except equipment. Changes larger than this must be justified in a way that will be understood by the reviewers.

SPECIAL FUNDING FOR TRAINEES

Underrepresented Group Supplements Represent "New
Money" to Encourage the Careers for Appropriate Individuals

As part of a commitment to encourage individuals who are members of underrepresented minority groups or who are disabled to enter research careers, the NIH allows investigators with two or more years of support remaining on their research grant to apply for an Underrepresented Group Supplement. These supplements fund the salary and some supplies for these trainees. These funds are in addition to those approved for the grant. The funds support an additional individual to work on the project.

Trainee Funding Is an Important Research Support
Mechanism and Helps Establish You as a Mentor

Equally important are applications for trainee support. These can be made directly to an individual trainee by the NIH, such as the National Research Service Award for postdoctoral fellows. Other trainee support goes to a department or institution to support a training program. For these, the trainee applies directly to the training program on campus.

Many Center Grants Have Funding for Junior Investigators

Some center grants (for example, the Mental Retardation Research Centers) and other mechanisms include small grants for junior faculty members, as well as providing them access to core resources. Not only is this funding and access helpful financially, but the additional mentoring and the opportunity to interact with senior faculty is invaluable. A halo effect may also operate when a such an awardee applies for their own grant.

INTERACTION WITH THE REVIEW
GROUP BEFORE REVIEW

If Your Grant Was Assigned to the Wrong
Review Group, Contact the SRA Immediately

The notification of assignment to a review group includes the name and contact information of the SRA assigned to the review group. If you are not assigned to the review group you requested, you should immediately contact the SRA of the review group to which you were assigned and the SRA of the review group you requested. You should then contact the CSR and copy the two SRAs. In this communication, indicate how your proposal is concerned with the topics of the preferred review group.

If Your Grant Was Assigned to the Correct Review
Group, Contact the SRA about the Format and
Deadline for a Supplemental Update

If you have been assigned to the review group and CIs that you requested in your letter of transmittal, you should immediately contact the SRA to determine the format and deadline for supplemental materials for your proposal. Since the SRAs are responsible for forwarding copies of the update for your proposal to the reviewers, they determine the length and due date of your update material. You will be awarded a grant based on what you accomplished as described in the Preliminary Results section of your proposal and in your update. If you are truly committed to the research program you have outlined in your proposal, you will have made progress between the time you submit your proposal and the SRA's deadline for supplemental material. In the absence of such progress, the reviewers will question your commitment and your ability to carry out the proposed research. You might argue that you should not be expected to have made progress in the absence of funding, but reviewers are not sensitive to this argument.

In your letter, you need to summarize the new data and update the status of publications. Remember this is an update, not a rehash of your proposal. Organize your letter using your specific aims. You will usually have four months between notification of your review group assignment and the due date for your supplemental materials (see table 6).

A Supplemental Update Demonstrates Your Ability to Make Progress

Lee and Stacey are new assistant professors in the same department. They both submitted a grant proposal with the same deadline. Lee's mentor encouraged her to submit a proposal update. She continued to work very hard in order to have new data for the supplemental material letter. Lee also revised a manuscript that was tentatively accepted after the original proposal was submitted. A letter of acceptance of the revised manuscript arrived before she sent in the supplemental materials. Stacey's mentor did not mention a proposal update. Stacey did not know about the possibility of submitting more material. Even though she accumulated more data,

*the review group was unaware of her progress. The primary and sec-
ondary reviewers of Lee's proposal were very impressed with her progress.
Lee was funded, while Stacey was not funded.*

HOW REVIEW GROUPS WORK

A Review Group Is Made Up of Regular and Ad Hoc
Reviewers and, Occasionally, Consultants

Your proposal will be assigned to a primary reviewer and a secondary
reviewer. They may be members of the review group, or may be ad
hoc reviewers, individuals with appropriate expertise who may be
under consideration for review group membership. They will attend
the review group meeting. Occasionally there is not the appropriate
expertise among the members or ad hoc reviewers in your review
group to evaluate your grant. In this case, one or more consultants
will be asked to review your proposal and prepare a commentary.
Consultants may not attend the review group meeting.

Not Everyone in the Review Group Will
Review Your Application in Detail

At the review group meeting, the primary reviewer will spend two
to five minutes presenting their opinion of your proposal. Usually,
the secondary reviewer will present only points of disagreement with
the primary reviewer. While the primary and secondary reviewers
are presenting their opinions, the rest of the review group may read
the abstract and leaf through your proposal. This reinforces the
importance of the abstract. Your abstract needs to be clear and must
include your specific aims and the significance of your proposed
research. Your abstract should not be written at the last minute. Your
proposal needs to be well organized, with headings for each specific
aim and subheadings for each subaim. This will facilitate consideration of your proposal during the discussion.

Table 6. *Timeline for Preparation of a Proposal Update*

	Month			
	1	*2*	*3*	*4*
Continue research	✓	✓	✓	✓
Prepare manuscripts	✓	✓	✓	✓
Submit manuscripts	✓	✓	✓	✓
Revise manuscripts	✓	✓	✓	✓
Prepare data for update			✓	✓
Write update				✓

The Primary and Secondary Reviewers
Generally Set the Range of Your Score

After the presentations of the primary and secondary reviewers, your proposal is scored. The range of scores is between 1.00 and 5.00 (often treated as 100–500). As in golf, the lower your score, the better. The primary and secondary reviewers give their scores first. Other members of the review group typically follow their lead. Occasionally, another member of the review group disagrees with the primary and secondary reviewers, or the primary and secondary reviewers disagree with each other. If there is a large discrepancy, there is a discussion including all members of the review group in an attempt to achieve consensus. If consensus is not reached, the members of the review group are instructed to "vote your conscience." The scores of all members of the review group are averaged.

Your Feedback Will Come in Written Form and May Be
Supplemented by Discussion with Your Program Officer

You will be able to access your score and the comments of the primary and secondary reviewers on eRACommons (https://commons.era

.nih.gov/commons/) after the review group meeting. The program officer from your primary CI will also have been at the review group meeting, so he or she may be able to help you interpret the comments of the review group. Written prior to the review group meeting, the reviewers' comments do not reflect the discussion and may not be reflective of your score. A brief section at the beginning of the review is intended to provide this information, but whether it does depends on the skill and conscientiousness of the SRA.

Your Percentile Rank Is More Important Than Your Raw Score

Some review groups are more rigorous than others. The scores from each review group for grants like R01s are rank ordered and combined with those from other review groups for calculation of a single percentile ranking for all of the proposals from all of the review groups. This is intended to correct the score inflation of some review groups. This final percentile is communicated to you and the CIs you requested for your proposal. Percentiles are like raw scores—the lower, the better. Some CIs have more money than others and thus are able to fund more proposals. Therefore, they can fund proposals at a higher percentile. If the first CI you requested is not able to fund proposals up to the percentile for your proposal, hopefully the second CI you requested will have more money, will fund to a higher percentile, and will be able to support your research.

If Your Proposal Is Triaged, You Should Seek Guidance in Deciding How to Proceed

Roughly half of all proposals are not scored by the review group. They are "triaged," meaning that they are not discussed during the review group meeting. You will receive the primary and secondary reviewers' comments, but you will not know where your proposal ranked. Were you at the 51st percentile or the 99th? Since your proposal was not discussed during the review group meeting, your program officer will be unable to provide any information in addition to the primary and secondary reviewers' comments. The program officer's experience may be helpful in interpreting the comments, and having heard the discussions of proposals that were scored, the program officer may be able to calibrate your proposal's relative position. By the current rules, you are able to submit one amended version of your proposal, but it is difficult to move a triaged proposal to a fundable score. The interpretations of the comments by the program officer and your mentors will help you know how to respond. The goal of the triage system is to permit more discussion of the better grant proposals, but it places a significant proportion of applicants at considerable disadvantage.

Do Not Give Up on Yourself

Chris and Lynn each submitted a grant for the same review cycle. Both grants were triaged, devastating both Chris and Lynn. They spent a lot of time consoling each other. Chris read the reviewers' comments carefully, talked to her mentors and program officer, and resubmitted at the next opportunity. Lynn was still licking her wounds, unable to focus on research or grant writing. Chris received funding for the amended proposal. The reviewers were impressed by her new data. Lynn still hadn't revised her proposal and had decreased her research effort.

CRITERIA FOR GRANT REVIEW

The members of the review group are charged with reviewing proposals in accord with the following criteria. Before submitting your proposal, you should consider each of these five points and emphasize them while writing your proposal.

I. Significance

While your research is important to you, it may not be significant in the larger context of the NIH. Too often, junior investigators are so focused on their day-to-day research that they cannot stand back to determine how their research fits into the discipline. While you may be pursuing a particular line of research in order to satisfy your scientific curiosity, this is not sufficient justification for the NIH to support your research. If you feel this is important basic research that will become more significant based on your proposed contributions, you will have to argue your position forcefully and effectively in your application. You need to show that your research will advance your field. If it will illuminate information, for example, about the pathogenesis of disease, then inject this as often as is reasonable throughout the Background and Significance section. Summarize this point in a paragraph at the end of that section. You should know that a single sentence or a brief paragraph making a glancing allusion to the significance of your work will not be sufficient.

II. Approach

Your Research Design and Methods section needs to provide the answers to the testable hypotheses derived from your specific aims. This section gives reviewers insight into your logical approach to problem solving. Your research design needs to include consideration of viable alternative hypotheses, approaches, and methodologies. To the extent possible, your methods should be standard, proven techniques, and if you have used them before, be sure to cite these references. If you are developing new methods, you need to include demonstration of their reliability and validity in your preliminary data.

III. Innovation

Your approach to your research should be novel. After a thorough literature search, you should propose specific aims that demonstrate your unique approach to this area of research. You may even challenge the status quo in the field if you have sufficient rationale, but beware that reviewers may be committed to the status quo. Therefore, in examining the roster of the review group to which your proposal was assigned, if any of the reviewers publish in this area, be sure to include references to their work. If you are challenging their approaches, you should work with your mentors to determine how to do this in the most sensitive manner.

IV. Investigator

Your qualifications to pursue this research are demonstrated throughout the grant. Your biosketch includes your training, experience, publications, and previous and current funding. If the grant application has a budget justification, you should develop this realistically to encourage the reviewers' confidence in your ability to accomplish your specific aims.

The review group members are also researchers. Their own experience and their review of numerous grants provide them with standard expectations for proposals. They readily spot any deviation from their expectations. Therefore, you should review successful applications from your mentors.

Your preliminary data show that you have already partially accomplished your specific aims. Novice applicants often think that if they put too much preliminary data in the proposal, they will not be awarded the grant. Reviewers want to approve grants that will be successful, and the more preliminary data, the higher the likelihood of success. Your choice of methods, anticipated results, and alternative approaches indicate that you can engage in critical problem solving to ensure completion of the grant. Your inclusion of collaborators suggests that you are able to utilize experts in the pursuit of your research.

Include a timeline at the end of your proposal. This demonstrates that you can plan and that you have the experience needed to recognize how long it will take to accomplish your specific aims.

V. Environment

The resources available for your use need to be carefully documented. On the Resources page, you need to detail the equipment you can use that is already available in your laboratory and core facilities. This demonstrates institutional support for your work. You should demonstrate a viable community of scientists ready to assist you with your research through your collaborators. If you are applying for a training grant, your mentor's scientific credibility and past trainee history are critical.

The Mentor's Skill in Mentoring
Must Be Sold to the Reviewers

Dr. Jones suggested that his postdoctoral fellow, Sam, prepare an application for an NIH individual National Research Service Award (NRSA). Part of the NRSA application called for a listing of Dr. Jones's previous pre- and postdoctoral trainees. When Dr. Jones completed this section, he was very strict in interpreting the guidelines and listed only those trainees who were in PhD training programs or who had completed MDs or PhDs and were in formal postdoctoral training programs. Sam's NRSA application was not funded. One of the criticisms was that Dr. Jones did not have sufficient mentoring experience. For Sam's amended application, Dr. Jones included less traditional trainees, such as master's degree students, medical students performing summer research projects, MD post-

doctoral fellows spending six months on a research project, visiting scientists coming for three to six months to do research, and an associate professor from another institution taking a one-year sabbatical with Dr. Jones.
Sam's amended NRSA proposal was funded. The reviewers had high praise
for Dr. Jones's mentoring abilities. The primary difference between the
original and the amended applications was that Dr. Jones carefully documented all previous mentoring experience.

RESPONDING TO THE REVIEWERS' COMMENTS

*If Unsuccessful with This Submission, Pay Attention
to the Reviewers' Comments, Since They
Will Guide Your Resubmission*

When you submit a grant proposal, you have put in a great deal of
work over a long period of time on a research program very important to you. If your proposal is not funded, it is normal to feel rejected,
angry, confused, depressed, disappointed, and just plain rotten. As
you read through the comments, you may feel that they could not
have read your proposal and say what they are saying. It is easy to
dismiss their comments as totally erroneous.

But the proposal was not funded. Usually each review has some
truth, or a misperception or misinterpretation that you did not prevent. You should put the comments in a drawer for a day or so until
you have calmed down and can read them more dispassionately. You
need to carefully evaluate how you can improve your resubmission
in line with the reviewers' comments. If you ignore their criticisms,
your proposal will most likely score worse in its next review.

The worst thing you can do is to call the SRA of your review
group and rant and rave about the unfair review your application
received. The second worse thing is to complain to the program officer of your CI.

Be Positive in Your Interactions with Your Program Officer

Kim worked for years to cultivate a relationship with the program officer of the CI most likely to fund her research. The program officer helped Kim decide which type of grant to apply for and answered her questions regarding each proposal. The program officer suggested that Kim apply for another grant in response to an RFA. When Kim received the review of her application for the RFA, Kim immediately called the program officer to complain and suggest that the problems with her proposal were due to the program officer's bad advice. Even though Kim called the next day to apologize, their relationship was never the same. When the program officer was selecting a special review panel in Kim's area of research, the program officer could not recommend Kim given her unprofessional conduct.

Use Your Network to Help You Craft Your Revision

When you have calmed down enough to engage in a rational discussion with your mentor, you should do so. You may also wish to seek the advice of other senior researchers in your area. These individuals will be able to advise you regarding the validity of the comments and recommend how you should respond to the review. You need to determine how to address the criticisms in order to improve the reviewers' perceptions of your work. You want to craft the revision of your application so as to sell it to them more effectively.

Your Mentors May be More Objective in Their Evaluation of Your Reviews

Sandy was devastated by the review of his proposal. His initial reading of the review was that he did not know how to formulate testable hypotheses, do a literature review, conduct research, or present a research plan. He was so embarrassed that he would not have shown Dr. Smith the review, but Dr. Smith asked whether he had received it. Dr. Smith's inter-pretation of the review was totally different from Sandy's. Dr. Smith pointed out that the reviewers suggested only the restructuring of his specific aims, the addition of one citation to his literature review, a possible second interpretation of one of his preliminary results, and two additional methods under the second specific aim. Noting that each of these suggestions was specific and constructive, Dr. Smith advised Sandy to incorporate them into the amended application along with additional preliminary data. Dr. Smith urged him to submit an amended application for the next possible deadline. The same information was there for Sandy and Dr. Smith to read in the review. Sandy took the criticism too personally. Dr. Smith's experience with grant proposals and their review enabled him to be more objective in the interpretation of the reviewers' comments.

Contact Your Program Officer If You and Your Mentors Are Concerned about Serious Flaws in Your Review

Flawed reviews can occur when one or more reviewers allow personal feelings to guide their review. If you find a large number of factual errors in your review and the tone of the review is malicious, you should discuss a possible appeal of your review with your mentors. Be sure that the review is indeed flawed. A difference of opinion is not a sufficient basis for an appeal. You do not want to get the reputation of being a difficult person. If you and your mentors decide that you have a flawed review, you should contact the program officer and determine whether they have any insight into the problem and request their guidance in proceeding with an appeal. You should write a letter and present point by point each item in the review that is incorrect or biased. Regardless of how upset you are, your letter will be more effective if you craft a careful and unemotional response. You must recognize that the appeal process frequently slows down your application process.

Seriously Flawed Reviews Can Be Appealed Successfully

Sam's mentor suggested that she apply for a postdoctoral fellowship grant. When the review came back, it was full of inaccuracies and was very vindictive. The inaccuracies included statements that her mentor had not produced academicians, even though her mentor had three former postdoctoral fellows and five former graduate students on faculty at leading institutions. A reviewer also claimed that Sam's mentor was no longer a productive researcher, even though he had NIH support and had ten peer-reviewed publications the previous year. Sam and her mentor wrote a letter documenting the inaccuracies and requesting another review. Upon review, Sam's postdoctoral fellowship was funded.

Do Not Give Up on Yourself

In nearly every case, your review will not be flawed. Regardless of how many grants you apply for, it is still devastating to be turned down. Too many applicants give up at this point. They turn over control of their research program to a review group. They forget that it takes time to learn how to write grant proposals. The writing skills for grant proposals are honed by responding to the critiques of reviewers.

Resubmit the Amended Application as Soon as You Can

You should have additional data and publications for this revision to show progress since the original application. Your responses to the opinions of the reviewers should be rational and based on the data. You should simply accept the proposed funding cuts and reduced duration recommended by the review group. There is no point in antagonizing the members of the review group. You should correct any factual mistakes in the review in a calm manner. The important suggestions in the review need to be dealt with in the introduction of your revised application. Any change from the original proposal needs to be italicized or underlined. Boldfaced changes are permitted, but they are hard to see when the proposal is copied. Have colleagues and mentors read the reviewers' comments and revised application to determine if they feel you have been as responsive as you can be.

Experienced Investigators Frequently Must Resubmit Applications

If you have a research grant, you should submit your competing renewal nine or more months before the end of your funding. If your competing renewal is not successful, you should request an unfunded extension of your grant. This allows you to carry an unspent balance forward to the next year so that you can continue your research while you prepare your amended application.

CLINICAL TRIALS RESEARCH

The principles described above are all relevant to clinical trials research proposals. The one difference between clinical trials proposals and those submitted to government agencies (e.g., NIH, state health departments) or foundations is that many of the costs for the clinical trials are negotiated. The goal of the drug company sponsoring the clinical trial is to keep the costs as low as possible. For the research institution participating in the clinical trial, the goal is to maximize the revenue. For the investigator at the research institution, the goal is to successfully complete the negotiation process so that their patients can participate in the trial. The investigator may have colleagues at the drug company with whom he or she has worked before, and may be influenced in the negotiations by those colleagues. We have seen cases in which the costs negotiated are too low but the institution is obligated by the contract to complete the trial. As a consequence, such trials that will bring the company's product to market end up being subsidized by the research institution and/or the department.

Most institutions and many departments have individuals designated to negotiate with the drug companies for clinical trials. You should take advantage of these individuals' expertise. They know the costs, including the hidden ones, of clinical trials, and they are skilled negotiators. We have found that the drug companies are quite willing to pay the appropriate costs when these are explained. The company needs the patients to achieve, for example, U.S. Food and Drug Administration (FDA) approval.

An Institution's Clinical Trials Office Can Make a Difference

Dr. Smith at University A and Dr. Jones at University B were in the same subspecialty and were friends; both decided to participate in Newco's trial of its proposed new drug. Dr. Smith had friends at Newco and worked with them to develop the proposal and the budget. Dr. Jones also had friends at Newco. However, Dr. Jones's university had a clinical trials office that would develop the budget based on the proposal's requirements and would negotiate that budget with Newco's representatives. At a meeting of the clinical trials investigators, Dr. Smith complained that he was funding a clinical nurse to supplement the research staff, which had been inadequate to do the trial. Dr. Jones was surprised, because she found this to be a well-funded trial. When they compared their budgets, Dr. Jones was getting 30 percent more than Dr. Smith for each patient enrolled.

CHAPTER 12

Developing Your
Teaching Skills

TEACHING IS THE CORE OF ACADEMICS

The common thread through all of academic life is teaching. While each academician develops their own style, everyone can be an effective teacher if they couple a sincere desire to effectively impart knowledge with sufficient time and effort. The motivating force for teaching is truly caring for and respecting students and trainees and wanting them to succeed. If you consider the qualities of the outstanding teachers you have had in the past, you can develop your own concept of an ideal teacher with attributes that you would like to incorporate into your own style as an educator.

THE TEACHER IS A GOOD LISTENER, ENTHUSIASTIC ENCOURAGER, AND EFFECTIVE MOTIVATOR

Truly caring for students and trainees translates into being a good listener and focusing full attention on the trainee. An effective teacher cares about students as people, beyond the educational situation. This caring does not imply total acceptance of all of the student's qualities. Instead, the teacher encourages positive behavior and specifies behaviors that should be improved. An effective teacher sets standards that motivate students to do their absolute best. Enthusiastic teachers who share with students a good sense of humor are also effective, by not letting the students take themselves too seriously.

Attempts to Prey on Your Sympathy

Dr. Johnson was in his first year of teaching at a small college. Two of his students were blind. Kelly had been blinded by an accident at age eighteen, and Kim had been blind since birth. The college provided student peers from each class to assist the blind students by reading the textbook and other materials to them, reviewing lecture material with them, preparing term papers as dictated by them, and giving the course exam questions and noting their answers. Kelly had two classes with Dr. Johnson. Fortunately the student peers for each of these classes came to Dr. Johnson and said that Kelly expected them to write the papers for the course, and to provide the answers for the exams, with no input from Kelly. This was in contrast to Kim, who was very independent and did not ask peer tutors to submit their work as Kim's. Dr. Johnson met with his department chair, and then they both met with Kelly to discuss the inappropriateness of Kelly's behavior.

THE TEACHER RESPECTS THE STUDENT

Respecting students and trainees requires the teacher's willingness to communicate through explanation or demonstration to clarify information. Effective teachers strive to answer questions efficaciously without making students or trainees feel inferior or guilty for asking questions rather than immediately grasping the presentation. Respect for students also requires that the teacher is predictable and consistent. The teacher should be humble enough to be willing to say "I don't know" if that is the case. Another aspect of humility comes into play if the trainee presents information suggesting the teacher has been incorrect: the teacher demonstrates respect for the trainee and intellectual integrity by admitting that he or she made a mistake.

Control of the Learning Environment Does Not Mean a Bully Pulpit for the Professor

Dr. Robinson was a senior professor who maintained a position because of a successful lawsuit. Dr. Robinson was very biased against members of a particular ethnocultural group. During class, Dr. Robinson would constantly berate members of this group. This made all of the students in the class uncomfortable. The students went to the dean and complained. Finally, the dean agreed that Dr. Robinson would no longer be teaching. Because of the lawsuit, Dr. Robinson continued to hold the position, but at least no more students had to suffer.

KNOWLEDGE BEGETS ORGANIZATION AND CONFIDENCE

Wanting students and trainees to succeed requires the teacher to communicate what that success will entail, that is, to have clear, explicit expectations that are effectively communicated to the trainee. It goes without saying that teachers should be thoroughly knowledgeable in their area of expertise. A strong knowledge base enables the teacher to be organized and to exude confidence. The knowledge needs to be organized so that students understand the logic of the presentation.

Lack of Organization Is Unfair to Students

Dr. Livingston was a tenured associate professor. Dr. Livingston repeatedly came to class late and occasionally canceled class for no apparent reason. Dr. Livingston was constantly changing the requirements for the course. After the first written assignment was turned in, Dr. Livingston switched from biweekly written assignments and monthly exams to one final paper at the end of the term and one final exam. The students did not receive the graded first assignment until after they had submitted their final paper. The course grades were more than a month late. This meant that those who expected to graduate at the end of the term were unable to do so. When they complained to the department chair, the students were told that Dr. Livingston had tenure and there was nothing the department chair could do.

THE TEACHER IS A LEARNER

You never learn so much as when you teach. Teaching encourages continued growth and professional development. In the course of explaining material you think you have mastered, students' and trainees' questions may be difficult to answer, causing you to present additional explanations or novel examples. You may even have to study and prepare further to deal with answering the questions at a later time.

BE RESPONSIVE TO YOUR CRITICS

While teaching evaluations may not always be flattering, they are always helpful. If you want to believe the glowing endorsements, you have to accept the grain of truth in the stinging condemnations. Use evaluations to become an even better teacher. You may not agree with each comment. If you see a pattern to the criticisms, however, you should definitely take heed and make adjustments to your organization and/or delivery. You should discuss this pattern of criticism and your responses with your mentor. You should ask your mentor and other senior faculty to attend one of your presentations to provide feedback. At most institutions, evaluation of your teaching by faculty peers is required for promotion and tenure decisions, so be proactive and responsive. You are competing with yourself (and not with your colleagues), which means you should attempt to improve your own teaching evaluations each year, no matter where your scores fall in comparison with your colleagues'. If you show this improvement in your own performance, you will continuously improve your teaching.

ESTABLISH MECHANISMS
FOR FEEDBACK

In addition to being self-critical, it is extremely important to take advantage of every reasonable opportunity to get meaningful feedback on your teaching. If there is no mechanism or requirement for teaching evaluations in your department or at your institution, develop a mechanism for yourself. In addition to student and colleague evaluations, follow your students' progress beyond your classes to see if you have provided them with a firm foundation for future learning. Course evaluations are like publications, patents, and funded grants. They are an irrefutable currency in promotion and tenure decisions.

TEACHING MAY NOT BE THE
ONLY CRITERION FOR PROMOTION

Never assume that superb teaching is "enough" for promotion. It is critical that you know the practical aspects of this process for your academic series (e.g., tenure, research, or clinical track). If creative productivity is required in your faculty series, you must be aware of the requirements regarding, for example, publications and grant funding. You should ask a mentor to assist you with promotion and tenure who can introduce you to the unwritten rules while explaining the written rules for this process at your institution.

Teaching can provide immediate gratification for the talented educator. Students respond to the effective and enjoyable presentation of information. However, in many institutions, teaching is necessary, but not sufficient, for promotion. You need to know the requirements for your institution and develop a disciplined plan of action for meeting them, including a timeline to achieve intermediate goals. You also need to utilize your teaching materials to develop review articles, chapters, and textbooks.

TEACH AT EVERY OPPORTUNITY—
AND PLAN SUCH OPPORTUNITIES

You should use every opportunity, both formal and informal, to teach. Anyone who asks a question and is ignored may not ask a question again. You will lose an important opportunity for teaching. Often, questions should be answered briefly with the possibility for more extensive discussion left open.

If students are doing independent research with you, you should schedule group and individual research meetings. This encourages trainees to ask questions and provides the opportunity for important discussions beyond the research. These additional discussions represent extremely valuable mentoring, often giving the student insight into what the mentor's work life is like.

LEARN TO DEAL POSITIVELY
WITH THE DISRUPTIVE STUDENT

Occasionally, you will have students who ask questions to enhance their self-esteem or to deliberately embarrass you. Both of these situations are disruptive and need to be dealt with effectively. In either case, you can deflect such questions by suggesting the student review the literature on the topic and make a presentation at the next meeting. Since the motivation for this type of student behavior has nothing to do with learning, giving the student more work is usually an effective countermeasure and enables you to keep the class on track.

KEEP APPROPRIATE DISTANCE
BETWEEN YOU AND YOUR STUDENTS

Some students may want to be friendly with you. This may be a particular problem if you are relatively close to your students in age. It may initially take the form of compliments, such as about your clothes, your hair, or your teaching skills. It may be expressed as a suggestion to meet for a cup of coffee.

Recognize that as a teacher you are judging the academic performance of your students. If it is perceived by your students and/or colleagues that some of your students have a more privileged relationship with you than others do, this will be considered behavior lacking in professionalism. Some students may claim that you have a conflict of interest when it comes to grading. If these concerns are brought to the attention of your superiors (e.g., department chair, dean), a reprimand could result that might become part of your personnel file.

There are many strategies to address overfriendly students and maintain the professional distance required of the objective teacher. For example, if a student compliments you, you may say something like "Thank you, but we are here to learn the subject." At the second offense, you should skip the thank-you and make it clear that you want to focus on the discussion of the subject. If there is a third offense, you should make it clear that this behavior is inappropriate and it must stop. After class, you should immediately contact your mentor and discuss why this is occurring and request additional strategies.

If a student invites you to meet outside the classroom (e.g., for coffee), you might suggest firmly that this would be inappropriate while the student is in your class. If the motivation is related to a desired favored position in the class, you will not be bothered by that student again. For the same reasons, never accept a gift from a student.

CONSIDER YOUR INTERACTIONS WITH THE MEDIA TO BE OPPORTUNITIES TO TEACH THE PUBLIC

As your career develops, you may be asked by the media to give interviews. These will come because of your expertise in a particular area. You should always go through the public relations group of your institution/company/agency. They will protect you from unprofessional interviews with undesirable publications. They will also use your interview as the basis for other media releases. They may offer to be present during the interview, particularly if there is concern that the reporter may be adversarial.

Before agreeing to do an interview, you should ask the reporter what their deadline is and who else is being interviewed. If you have time, it would be helpful to review the reporter's other work to get a sense of their style. Prepare for the interview by focusing on your key messages in language that is understandable by the public. You want to teach the public through the interview. You should not be passive. Seize every opportunity to present your message. If the reporter is negative, turn the question around so you can articulate your positive message. Be brief and state your message clearly. If you can cast your message in a memorable sound byte, this will increase the likelihood that you will be quoted. You should also correct any errors. You should ask to review any written material the reporter produces before their deadline, though many print media journalists will not permit this.

In the interview it is OK to admit you don't know everything. You can refer the reporter to another colleague or offer to get back to them with the information they want.

Remember you are being interviewed the whole time. There is no such thing as "off the record." Also beware of the "friendly chat" after it appears your microphone has been turned off. It may not have been turned off, or the reporter may be recording from his or her microphone. The interview is not over until the reporter and the entire crew have left the room.

If your institution/agency/company offers media training, take advantage of the opportunity. You will find that media training helps you answer questions in other contexts, such as teaching.

CHAPTER 13

Developing Your Leadership Skills

YOU ARE A LEADER

Because of your selection of an academic career, you have chosen to be a leader. You have already assumed a number of leadership roles: as a teacher, as a researcher, as a professional, as a member of professional organizations. Your continued participation in the academic infrastructure is essential to the future of your discipline and the future careers of your mentees.

VALID REASONS TO BECOME A LEADER THAT ARE REALISTIC AND ORGANIZATION-CENTERED

Leaders may have a new vision for the group and want to implement it. They may want to move their field in a new direction to anticipate future needs. They should have a thorough understanding of where the group is at the present time and where they would like it to be in the future. With this knowledge, they can advance the group agenda to achieve outcomes consistent with the leader's own agenda.

Leaders want to engage in creative problem solving. They can develop a model program to test their ideas in order to determine what does work and what does not. They may want to reconcile differences between two or more groups to move them ahead toward shared goals. They may want to redeem a group that is struggling to survive and yet has important goals. They may want to obtain recognition of a group's achievements that have gone largely unnoticed.

They may want to have a larger impact as a mentor and help to develop the careers of others. They want to support the academic infrastructure that supported their careers.

INVALID REASONS TO BECOME A LEADER THAT ARE UNREALISTIC AND SELF-CENTERED

Some individuals seek leadership positions for power—to increase the number of individuals under their control. They want payback—to reward their friends and punish their enemies. It is inappropriate to use the group to advance your personal goals. True leadership requires shared governance—group members striving together to reach a common goal. Leadership requires vision, self-sacrifice, and hard work.

Others aspire to leadership roles to enhance their self-esteem. Leadership means leading by example. No leader should ask someone to do something the leader has not done or is not willing to do. A leader should always be willing to do more than any group member.

Some see leadership as an honorific right—the natural next step in the academic structure. They see leadership as an acknowledgment or a reward for longevity. Certainly one has to survive in order to become a leader, but survival is not enough. Demonstration of leadership skills and leadership qualities is required. Some consider leadership to be a birthright: "my mentor/relative/classmate was/is a leader, so I should be a leader by virtue of our shared relationship."

Simple association with a leader is not sufficient to convey the qualities and skills required for leadership.

Some will say, "I know better than anyone else" or "I'm smarter than anyone else." Certainly ability is important, but intellectual ability does not equal leadership ability or skill.

If one is bored with one's current position, one might view leadership as change for the sake of change. Responsible leadership, however, involves too much work to be considered solely as providing variety.

Leaders Must Have Appropriate Preparation to Establish Leadership Skills

Kim is a full professor and a successful researcher. She is focused on her research, to the exclusion of other aspects of her career, steadfastly refusing to serve on committees, take on extra teaching responsibilities, or pursue additional service commitments. Looking for a new challenge, Kim decides to become a department chair. She meets with her current department chair to discuss her new goal. The department chair is caught totally off guard by her plans. The chair tries to explain that one becomes a department chair by demonstrating leadership in a variety of areas, including research, administration, teaching, and service. Since Kim has always greatly frustrated the department chair by refusing to participate in the departmental infrastructure, the chair has a difficult time reconciling her past behavior with her new goal. The department chair suggests that Kim participate more fully in the department in order to test the reality of her new goal.

EVEN IN GROUPS OF TWO INDIVIDUALS, ONE WILL BE A LEADER

Any productive group of two or more individuals involves leadership to keep the group moving toward a common goal. With two individuals, the leadership may alternate, depending on the nature of the issue being addressed. With three or more people, there may be layers of leadership.

To train mentors, we have a senior mentor work with a senior trainee, or a young faculty member in mentoring a more junior trainee. This "mentoring the mentor" model not only helps assure an optimal training environment but also provides credibility for an NIH-funded training program.

IMPORTANT QUALITIES IN SUCCESSFUL LEADERS

A successful leader shares his or her vision with the group. A leader trusts others and is worthy of their trust. You should forget the power pyramid if you wish to be effective. A successful leader encourages full participation by group members. A leader has personal integrity, is a good mentor, respects diversity, encourages creativity, is a good listener, and is well organized. The successful leader understands the needs of group members and works to meet these needs. A good leader is able to admit a mistake, a lack of knowledge, or a lack of ability and is willing to ask for help. A leader is fearless—willing to make a decision after assessing the facts and live with the consequences. Decisiveness is essential to move the group forward.

Realizing Your Vision as a Leader

An effective leader needs to be able to formulate a vision that will serve to move the group, and perhaps the discipline, forward. Such a leader should provide a plan to move the group toward long-term goals that inspire collaboration among group members. The leader needs to articulate short-term goals to achieve their vision. The leader needs to be an effective communicator to excite group members about their vision. The leader uses shared governance to encourage group efforts in support of their vision.

Personal Integrity Is One of the Most
Important Qualities of a Leader

Leaders must have integrity, which includes an inner strength of character that enables them to be fair, consistent, highly moral, and honest with themselves and others. They are able to stand back from the clamoring of individual group members and make an independent decision based on what is best for the whole group. As part of their personal integrity, leaders treat each group member with respect and consistency. They maintain the higher moral ground in the face of temptation to make an easier, sometimes more popular choice.

Leaders Respect Requests for Confidentiality

The leader commits to keeping information confidential when requested to do so. They refuse to divulge a confidence in the face of repeated attempts to force revelation. The leader remains above gossip and rumor at all times—that is, does not tell anything held in confidence and likewise does not listen to what others say with the intent of passing on those murmurings as a false currency of power.

A Leader Develops Effective Listening
and Communication Skills

A leader is able to express an idea and to listen to feedback from group members regarding the idea. The leader focuses on the person speaking and asks appropriate questions to clarify points. Leaders are willing to accept criticism from others to refine an idea that they feel is important to the group's progress. By improving the idea with input from the group and through effective use of their communication skills, the leader can sell the idea to the group.

Leaders Accept Their Limitations and Seek Help from Others

Leaders recognize that no one is good at everything. They know when they need help and actively seek such help. They assemble a team of individuals with complementary skills. This team may include individuals who will critically evaluate the idea or task at hand, as long as the criticism is constructive and not destructive.

Leaders Show Humility

Leaders realize that any one person has a small role to play, including the leader. They choose to celebrate the success of others rather than their own. They share credit with the whole group for reaching their goals. They are willing to perform any task, no matter how trivial or how difficult.

Leading Is Done by Example

Leaders are willing to work harder, faster, and smarter than any other group member. They accept group decisions graciously and follow through. They are willing to place their own well-being at jeopardy to benefit the group. This does not mean that they never compromise, but that they have a clear set of guiding principles and know that to compromise such core values is destructive to the group's mission.

KNOW WHEN TO SAY WHEN

A successful leader knows when to say when. If some leadership is good, is more necessarily better? In other words, there may be a limit to the number of leadership roles that an individual can take on and remain effective.

You should know yourself. You need to consider the implications of a new leadership role for those who are important to you in life.

The mindless pursuit of ever-increasing leadership roles is bound to lead to disaster. The successful leader knows when they have reached their comfort zone, and should choose to thrive at that leadership level.

Compare what you do now to the leadership role you are contemplating. You can set up your own version of vocational choice tests:

- Would you rather give a lecture or attend a committee meeting?
- Would you rather edit or write a manuscript?
- Would you rather be on the planning committee for a professional meeting or work on your own research?
- What would your typical day be like? Ask those currently in similar roles what they like about what they do.
- Who will your boss(es) be? Do they have personal integrity? Can they support your vision? Are there enough financial resources available to support your vision? You need a copy of the organizational chart to understand the reporting structure. Are you responsible to a board in addition to your immediate boss?
- What is the trajectory for this organization? Is the organization on the upswing or the downslide? If the organization is on the downslide, can implementation of your vision reverse the trend? Is the organization able to change?
- How do you feel about the members of the group you would lead? Do you respect them? Can they learn to respect you? Can you work well together? Will they respond to your vision? If there are other candidates for the position, what do you know about them and what does that tell you about the organization? Can you be passionate about the position?
- Does the organization have the will to move forward, or are its members wedded to the status quo?

Consider the implications of a new leadership role on those who are important to you:

- Do you have to move? What will be the impact on your family, trainees, and colleagues?
- Will you have less time for your research? How will this impact your trainees and colleagues?
- Should you cut back on your teaching in this new position? What will be the impact on your colleagues?
- Will you have to give up other leadership responsibilities? How will this impact those organizations?

Before committing to proceed toward a leadership role, you need to determine if you have a vision for what you would want to accomplish in that role. Without a vision, you will be hard-pressed to sell yourself as a serious candidate for the job. Does your lack of vision mean that you would lack commitment to the position, or that you do not fully understand that position? Without a vision and well-defined goals leading to its achievement, it will be difficult for you to gauge your success.

WHAT ARE THE OPPORTUNITIES FOR LEADERSHIP IN RESEARCH?

The first step toward leadership in research is to establish your own research group and your independence as an investigator. Another leadership role is serving as the principal investigator of a research grant. You may then progress to being the principal investigator of a program project or center grant. Your research role may eventually expand to being the director of a research institute.

Publishing is another leadership path. As a successful author of peer-reviewed publications, you will be asked to review manuscripts. Thorough, thoughtful reviewers are asked to join journal editorial boards. This could eventually lead to being editor-in-chief. In addition, your publication leadership may lead to requests for you to prepare mini-reviews or book chapters on your research topic. If these are successful, you may be asked to edit or co-edit, or to author or co-author, a book.

WHAT ARE THE OPPORTUNITIES
FOR LEADERSHIP IN TEACHING?

Teaching is definitely a leadership activity. Successful teachers develop new courses, new course models, and new curricula. They may be asked to write or co-write textbooks. They may have increasing responsibility in the academic administration.

WHAT ARE THE OPPORTUNITIES FOR
LEADERSHIP IN ADMINISTRATION?

Whether you are in an academic institution, company, agency, or professional organization, opportunities exist for promotion with increasing leadership responsibilities. You can begin as a committee member and move up to chair the committee. You should choose the committee based on workload and the committee's effectiveness. Decision-making committees should be given precedence over committees that exist solely for the leadership to dole out information. You should also determine whether you will be able to make a difference as a committee member. Committee service can lead to defined leadership roles within the entity. If you are offered committee leadership opportunities, consider the reporting structure and political risk and your ability to influence policy.

As you assume increasing leadership roles, you may have to end some or all of your academic duties. Before you accept a role, determine whether you will be able to have an impact or whether your position will be largely ceremonial.

If you are considering running for an office in a professional organization, remember that you will be representing the interests of the organization, not your personal interests. You need to assess the status of the organization—for example, its political position in the professional hierarchy and its financial status.

STEPS IN DEVELOPING LEADERSHIP SKILLS

To develop your leadership skills, you should seek out mentoring by leaders you respect. You should share your interest in leadership with current and past leaders. Mentors can provide advice, examples, introductions, information about organizational protocol, and recommendations or nominations for experiences that will prepare you for leadership roles. If you can't follow the recommendations of your mentors, you should discuss this with them and describe your reasoning.

You should obtain feedback from your mentors regarding your performance. You should strive to improve based on their recommendations. Instead of asking "Is this good for me?" start asking "Am I good for the group?"

You should try different leadership activities, such as the following:

· Organize a session at a meeting or a whole meeting
· Develop a new program for the group
· Represent the group at meetings of other organizations with similar goals
· Encourage others to join the group
· Apply for a grant to support the group's activities
· Nominate someone for a leadership role or an honor

You can also read self-help books and attend leadership workshops. You will have new perspectives after studying examples of leadership outside your own field. You should analyze your behavior in leadership roles. You should be self-critical and determine what is effective and what isn't. You should compare and contrast different situations to extract general principles, and consider how group members react to your strategies.

You should be careful to modify your behavior based on suggestions from group members. Its important to consider input from a variety of group members and not to capitulate to the vocal minority of complainers when the silent majority seems content with the status quo. You should develop creative, nonjudgmental mechanisms to get feedback from all group members.

You need to make sure that your leadership style is internally consistent. We all like to be able to predict the behavior of others, especially leaders. You should not talk of shared governance if you are unwilling to allow discussion of a single issue. You should not expect more of group members than you expect of yourself. You need to evaluate strategies for implementing your vision, making sure the strategies don't become ends in themselves.

You should volunteer and/or accept opportunities that are important to your leadership goals or are offered by someone you can't turn down. You should be willing to perform "menial" tasks. You need to learn how the group functions from the ground up. Some leadership roles are reserved for those who rise up through the ranks. Accepting difficult tasks and accomplishing these with grace, finesse, and timeliness will earn you the group's gratitude and respect. This is important when nominations for leadership roles are being considered. You should express your willingness to accept increasing responsibility if this is consistent with your ability to accomplish the work. You should volunteer to accept more than your "fair share" of the workload if you will be able to carry it out.

Take a chance and run for an office even if you don't think you are likely to win. You will improve your name recognition and increase the likelihood of being considered for other openings. You will earn the gratitude of the members of the nominating committee. You may surprise yourself—you might actually win.

BURDENS AND BENEFITS OF LEADERSHIP

As a leader, you will often sacrifice yourself for the group. You certainly will be committing time that you may have planned to use differently. You may have to sacrifice some of your personal goals in order to meet the goals of the group. In fulfilling your leadership role, you may have to contribute more effort than the other group members.

Leaders must work hard to maintain the high moral ground. It is not always easy to maintain confidences or to be consistently fair. It is hard not to show favoritism to group members.

You must know how to delegate work and authority. It is hard to help group members do something themselves when it would be easy for you to do it yourself. But this delegation is part of your role in mentoring future leaders. It is also important for shared governance in the group.

It is difficult but necessary for a leader to be self-critical. You need to work to change for the better. You have to be humble enough to accept help from others and acknowledge when they perform better than you could.

It is often difficult to maintain a positive attitude in the face of adversity. Regardless of how you feel personally, you need to be a cheerleader for the group. It is hard to give bad news to good people and to expect them to take that information and move forward. But you have to learn how to motivate the individuals and the group even in difficult times. This is when it is important to have a well-articulated vision that you are moving toward, even if it appears the movement is slow.

On the other hand, you may have the satisfaction of achieving your vision more quickly through group effort. Mentoring others to become future leaders brings continuing rewards as they take on leadership roles. It is gratifying that you can be a role model for others.

As a leader, you should look at the changes in the group under your leadership. You should determine progress in establishing shared governance, the impact of the group's activity, improvement in morale, and movement toward your vision.

You will experience personal growth as a leader. You will learn to be self-critical. You will learn how to set long-term and short-term goals for the group and yourself. You will develop the ability to inspire others with your vision.

IN ACADEMICS YOU OWN YOUR OWN CAREER

You need to develop a vision for your own career with the same kinds of short-term and long-term goals you would devise for a group. You need to review your progress on a regular basis.

Recognize that you will reinvent yourself many times throughout your career, if you are to be successful. Business organizations know that, if they wish to keep an individual in the organization, they must create a new position for them every six to ten years. You need to understand this in your own career. If you become bored and unproductive, you will become disappointed in yourself. Rather than allowing that to happen, be proactive and examine your opportunities. Consider taking on a new leadership position as part of your reinvention.

CHAPTER 14

Conflicts of Interest

Conflicts of interest arise when your decision making could be influenced by a selfish motive, in addition to or in opposition to the facts. For example, if you are asked to evaluate the research of your cherished mentee or your valued faculty colleague, your feelings for them could cloud your judgment. In either situation, a favorable outcome would also benefit you.

The best way to deal with a potential conflict of interest is to recognize it and to acknowledge it. In some cases, acknowledging your conflict of interest will mean that you must deal with it by recusing yourself from the review process. At other times, you need to inform others and let them draw their own conclusions. If you have any doubts, it is wise to err on the side of caution.

AS A MENTOR

As a mentor, you should proceed with the highest level of integrity in advising your mentee. If your mentee is exceptional, your self-interest would dictate that you would maintain the relationship indefinitely for your own benefit. As a true mentor, you would provide opportunities for skill development leading to the mentee's true independence as expeditiously as possible. A conflict arises when a mentor keeps an outstanding mentee in the mentor's research group longer than is appropriate in order to benefit the mentor.

Time to Move On

Kelly was pleased to be selected for a postdoctoral fellowship with Dr. Clark, since Dr. Clark had just received a five-year grant that would support Kelly the whole time. Kelly was a very productive researcher, had two publications the first year, helped Dr. Clark write a research grant, and taught one of Dr. Clark's courses. By the end of the second year of the fellowship, another faculty member in the department suggested that Kelly apply for faculty positions, noting that sometimes it took more than a year to get a good job and Kelly had already accomplished a great deal. When Kelly approached Dr. Clark with this idea, Dr. Clark rejected it immediately. Dr. Clark felt that Kelly had a lot more to learn and shouldn't waste time on applications. When Kelly asked Dr. Clark what was going to be different for years three to five of the postdoc, Dr. Clark was at a loss, since Kelly was so exceptional. Kelly pursued the job applications and was hired for the next fall. Dr. Clark remained resentful that Kelly left early, feeling that no one would work as hard or as effectively as Kelly. Dr. Clark was interested only in the benefits of Kelly's hard work, not in helping Kelly take the next career step.

AS A MEMBER OF THE ADMISSIONS COMMITTEE FOR A GRADUATE OR PROFESSIONAL SCHOOL

You may have the best intentions as a member of an admissions committee. However, sometimes your colleagues or other advocates for an applicant may overstep their bounds and put pressure on you to ignore the due process of review in favor of their applicant. While colleagues can provide letters of recommendations, their contribution to the process ends there. If it is not appropriate for certain advocates (e.g., family members) to provide letters of recommendation, because of a real or perceived conflict of interest, they should have no role in the process at all.

If there is an applicant with whom you have a personal history or connection, you must recuse yourself from review of their application and acknowledge why you are doing so to your colleagues. You can provide a letter of recommendation for this individual if you explicitly state your relationship to the applicant. A letter from someone the committee knows and respects carries more weight than a letter from someone they don't know or from someone they do know but don't respect.

If you learn something about the review of this individual, or about the decision regarding their application, you cannot provide them with this knowledge. Information needs to flow through the usual committee channels in the form of an official offer of admission or a denial.

AS A REVIEWER OF ABSTRACTS
FOR A PROFESSIONAL MEETING

When you are reviewing abstracts for a professional meeting, the topics may be within your broad area of research interest. Occasionally, an abstract might be from a direct competitor of yours. If this is the case, you need to judge the abstract on a purely scientific basis. If an abstract author was a trainee, collaborator, or colleague at your institution within the past five years, you need to acknowledge your conflict of interest and recuse yourself from the review. Other members of the program committee will review this abstract, so it will be scored.

AS A REVIEWER OF MANUSCRIPTS
FOR A PROFESSIONAL JOURNAL

If you are asked to review a manuscript whose author is a direct competitor of yours, you deal with any potential conflict of interest by agreeing to review immediately, by reviewing on a purely scientific basis, and by submitting the review on the same day you agreed to review. It would be a conflict of interest to slow the review of such a manuscript by not agreeing to review immediately and by not submitting the review immediately. You are the best-qualified person to do this review, so you should do it expeditiously. Editors keep track of the time to agree and the time to complete the review. They have a commitment to the authors and should not allow any subterfuge.

It is very frustrating for editors when a potential reviewer says they are competitors and therefore won't review. The editor knows that the potential reviewer is in the field, which is why they were invited in the first place. Who knows the field better?

As editors, we have noticed over the years that individuals who are the most demanding regarding the timeliness of reviews of their manuscripts are the least likely to agree to review the manuscripts of others. In their minds, it is better to receive reviews than to give them. We would suggest that their selfishness is a conflict of interest.

If everyone behaved the way they do, there would be no peer review of manuscripts. If you write, you must review. Our only asset is time. By refusing to review in order to have more time to devote to their own work, these authors are avoiding their responsibility to support the infrastructure of science. What they don't recognize is that editors have long memories for individuals who don't review. When it comes time to add members to the editorial board, these individuals will not be included, because they haven't met the criteria of both publishing and reviewing for the journal.

An Editorial Conflict of Interest

Dr. Jones is the editor of a journal. Dr. Johnson submits a manuscript to Dr. Jones's journal on a topic that Dr. Jones's colleague Dr. Smith is working on. Dr. Jones immediately contacts Dr. Smith and tells Dr. Smith about Dr. Johnson's manuscript. Dr. Smith asks Dr. Jones to slow the review

*process of Dr. Johnson's manuscript so Dr. Smith can finish the research
and prepare a manuscript for the journal. Dr. Jones agrees to do so. Dr.
Jones's conflict of interest is seen in telling Dr. Smith about Dr. Johnson's
manuscript and in slowing the review of Dr. Johnson's manuscript so
Dr. Smith's and Dr. Johnson's manuscripts are published back-to-back
in the same issue of Dr. Jones's journal.*

AS A MEMBER OF THE SELECTION COMMITTEE FOR POSTGRADUATE OR POSTPROFESSIONAL TRAINEES

In this situation, conflicts of interest can occur if one member of the committee asks another member to support their candidate and offers to reciprocate. This is especially difficult when there are a limited number of training positions and a significantly larger number of applicants. These kinds of deals have no place in the selection process.

As in earlier examples, you must recuse yourself if you have a previous relationship with the applicant.

If these training situations involve multiple programs, departments, or campuses, there can be an inherent conflict of interest in sequestering all of the trainees in one particular area. This strategy may temporarily benefit the faculty in this area, but in the long run this type of decision making dooms the training program that was funded on the basis of collaboration among multiple programs, departments, or campuses.

AS A MEMBER OF THE SELECTION COMMITTEE FOR A NEW FACULTY MEMBER

Each member of the selection committee has an awesome responsibility to the applicants. When it comes to hiring a faculty member, it can be easy to have a conflict of interest. You might want to hire someone who could teach one or more of the courses you teach, thereby relieving you of this duty. You might want to hire someone who has research skills that could complement your own. You might want to hire someone who would take on committee responsibilities you want to be rid of. You might want to hire someone you think you could manipulate. Rather than the self-interest leading to a conflict of interest, it would be better to consider candidates based on their potential contribution to the department, the institution, and the discipline.

AS A REVIEWER OF GRANT APPLICATIONS

In reviewing grants, you must recuse yourself from reviewing those of anyone who has been a trainee, collaborator, or colleague of yours in the past five years. If you are reviewing a competitor, you must do so solely on the basis of the science. Other members of the committee know that you are a competitor and will be watching your performance. If there is a significant discrepancy between your score and that of the other reviewer(s), this will be attributed to your conflict of interest.

AS A TEACHER

You need to treat all of your students the same. Some may try to be your friend, others may be particularly engaged with the material, and others may have tremendous potential. They all must be treated the same. This is especially important when it comes to grading. Grading should be based on a defined rubric established at the beginning of the term. Feedback should be as specific as possible to encourage improvement.

One obvious conflict of interest is the teacher who tries to be the students' friend. This teacher goes out of their way to solicit approval from trainees and to lavish praise (deserved or not) upon the students. Average grades may be very high compared to other professors. The students who don't play along and refuse to provide slavish devotion will receive lower grades. This entire paradigm is based on the teacher's insecurity. Trainees don't necessarily learn and are not held accountable for their behavior.

Trading Sexual Favors for Grades

Dr. Jones would select students with particularly desirable characteristics at the beginning of each term. These students would be offered private tutoring in Dr. Jones's office to improve their grade. Private tutoring consisted of the student providing sexual favors for Dr. Jones and "earning As." Those who refused to play along, or students who were not worthy of Dr. Jones's attention, received lower grades. Dr. Jones is ignoring institutional policy that explicitly forbids sexual relationships between students and their professors.

AS A LEADER

It is amazing how many individuals seek positions of authority to wield power by imposing their will on group members. This is not leadership. A leader is committed to improving the status of the group members, both individually and collectively. Leadership is the opportunity to mentor on a larger scale.

INSTITUTIONAL CONFLICT-OF-INTEREST POLICY

Each institution has its own policy regarding the declaration of potential conflicts of interest. For example, you must disclose potential conflicts of interest if you are performing human subjects research. If you are involved in the care of the potential participant, you need to acknowledge your potential conflict of interest as both clinician and investigator. If a research grant is supporting the research, this has to be acknowledged. If the research involves a drug or device and you received financial support from the manufacturer, the potential participants need to know about this. If you have an intellectual property interest in the drug or device, this also must be disclosed. The institutional committee on conflicts of interest may determine that more than disclosure is required.

Financial Conflict of Interest Not Disclosed to Participants in Clinical Trial

When one of the participants in a clinical trial died as the result of the treatment, the financial interest of one of the investigators in the treatment was revealed. The mother of the participant who died was very upset. She was concerned that the investigator wanted to make more money, while her daughter had volunteered to improve the care of other patients with her disease. The Food and Drug Administration forbade this researcher from engaging in any human clinical trials.

POLICIES OF PROFESSIONAL SOCIETIES AND OTHER LECTURE VENUES

Most professional societies require that you acknowledge any potential conflicts of interest before you begin your oral or poster presentation at the conference. This allows your audience to judge your contribution on this basis. They are particularly concerned about grants or other funding that you received from the manufacturers of drugs or devices.

In the past, a manufacturer would fund meetings, including travel, hotel, meals, and publication, so that lecturers selected by the company could present information regarding the company's product(s). This is no longer permitted. Instead, companies may contribute toward an educational activity or meeting, but may not control who is invited to present. It must be an unrestricted educational grant.

The invitations are made by an independent group of professionals who do not have any financial ties to the company.

Many institutions have policies prohibiting gifts or food provided by companies in an attempt to influence policy decisions made by the recipients. While those receiving these gifts or meals say they are not influenced, the data indicate otherwise. In addition, the public most definitely considers such material exchange provided by companies to be conflicts of interest.

CHAPTER 15

Human Subjects Research

Anytime you perform research with human subjects with the intent to publish your findings, you need to apply to your institution's, company's, or agency's review board. If your employer does not have an internal board, you can pay for the services of a board for hire. The purpose of this review is to insure that your procedures do not place your human subjects at unnecessary risk and that the participants in your study are fully informed regarding the purpose, methods, and possible outcomes of the research. Before undertaking research with human subjects, you need to complete the course(s) required by your review board.

Your application to perform human subjects research will include a protocol describing your proposed study and an informed-consent form. Some research does not require informed consent by the participant, but it is up to the review board to make this determination. The researcher cannot simply assume that the study is exempt from informed consent.

Storage of Patient Samples for Eventual Research

Dr. Smith was a very busy cardiologist who was on the faculty of a major medical school that performed a lot of heart transplants. Dr. Smith realized that samples from the transplant patients might be important for research on the genetics of heart disease, so every transplant patient's blood, own heart tissue, and a small portion of the donor heart were stored in Dr. Smith's freezer. Dr. Smith didn't feel the need to apply to the institutional review board for permission to perform this research. Dr. Smith didn't feel the need to provide informed consent to the transplant patients before their samples were taken and stored. As genes were cloned for different cardiac disorders, Dr. Smith planned to offer the stored blood and tissue to the researchers to obtain information about the patient's disorder. Dr. Smith planned on telling the patients the results of the research. There were very many issues that Dr. Smith had not considered that the review board would have raised: Did the patients want to participate in research? Did the patients want to know the results of the research? Given that Dr. Smith was not a board-certified medical geneticist, how could the research results be presented in the context of genetic counseling? How could Dr. Smith know that the results of the research were clinically valid? The patient samples were a valuable resource, but how would Dr. Smith determine which research was worthy of the use of the samples? Although Dr. Smith's intentions were well meaning, this is clearly a disservice to the patients. Dr. Smith needs to apply to his review board for approval to conduct human subjects research. The review board will decide if Dr. Smith can use the data already collected; if Dr. Smith needs to destroy all of the samples collected so far; if Dr. Smith needs informed consent from the patients; and, if Dr. Smith is permitted to continue, what changes need to be made.

CRITERIA FOR STUDIES EXEMPT FROM REVIEW

The following types of research may be exempt from consideration by the review board: education; publicly available material that does not identify the subject; government agency procedures; and food quality. However, if any such research is to be performed with prisoners, pregnant women, human fetuses, or neonates, prior review board approval is required.

When the institutional review board is determining whether or not a protocol is exempt from review, the following factors are considered:

Does the research pose no more than minimal risk to the subjects?

Are the subjects' privacy interests protected?

Will any private identifiable information be kept confidential?

What are the anticipated benefits to participants and others of this research?

What is the importance of the knowledge that will come from this research?

How will subjects be recruited and selected?

What informed-consent process will be employed?

The initial outcomes of the consideration of a study submitted for exemption could be one of the following: Certification of Exemption; additional information requested before a decision can be made; or exemption denied and applicant asked to submit a full application to the review board.

If a study has been deemed exempt by the review board, it does not have to undergo continuing review by the board. However, if the investigator wants to modify the exempt protocol, a modified application must be submitted and certified exempt before the revised study can proceed.

SUBMITTING A PROTOCOL TO THE REVIEW BOARD THAT IS NOT ELIGIBLE FOR AN EXEMPTION

When one is submitting an application for human subjects research for review by the board, the following materials are included:

Protocol

Health Insurance Portability and Accountability Act (HIPAA) application if access to personal health information is requested

Proxy consent application if some or all participants are expected to be unable to provide informed consent

Recruitment materials if necessary

Informed consent (ages thirteen years and up) and assent (ages seven to twelve years) forms

Surveys, questionnaires, and other testing materials

Sponsor's research protocol if supported by a drug or device company

Investigator's drug brochure if studying a new drug

Device brochure if studying a new device

Copies of the grant application(s) supporting the research

If this is a Health and Human Services multicenter clinical trial, copies of the approved sample informed consent and protocol

Having a protocol accepted by the review board on the initial submission is about as rare as having a manuscript accepted for publication without any changes. The researcher should make all acceptable changes as soon as possible. If there are changes the investigator does not want to make, the lack of change has to be carefully justified. The researcher needs to be sure that this refusal to accept the review board's suggestions is justified, and needs to explain the rationale as carefully and clearly as possible. Often this review process can drag on interminably. The investigator might find it helpful to consult with the staff or the chair of the review board. If these issues continue to be impossible to resolve, the researcher may be asked to appear before the review board. This is a very intimidating meeting for the investigator and should be avoided if at all possible. Often these issues arise from simple misunderstandings that should be resolved quickly in the interest of all concerned.

Once the protocol has been approved, the investigator submits an annual continuation application and consent and assent documents.

REPORTING AN ADVERSE EVENT
TO THE REVIEW BOARD

The review board is concerned with the safety and well-being of human subjects. The investigator agrees to report any adverse event immediately to the review board, to include an assessment of whether the adverse event was due to the study, and to describe any changes in the protocol to minimize risk and any alterations in the consent procedures and documents to inform participants of any new information.

The adverse event is reviewed to determine if any further action is required, if there are questions for the investigator, and if the entire review board needs to consider the adverse event. The review board will make one or more of the following determinations: the protocol needs to be changed to decrease risk; the informed-consent document needs to be changed to better inform participants of a new risk; the review period for the study may be shortened; current and past subjects may be notified of the new risk; the research will be stopped.

RESEARCH INVOLVING MEMBERS
OF VULNERABLE GROUPS

The following are considered vulnerable individuals:

Prisoners

Pregnant women

Fetuses

Neonates

Children

Mentally disabled persons

Physically disabled persons

Economically or educationally disadvantaged persons

Elderly persons

Terminally ill persons

Students at the institution

Employees of the institution

The review board does not want these individuals to be coerced into participating in research. The investigator may have to follow specific procedures to prevent coercion. There is also the question of whether research with these individuals should be undertaken given the risk.

Prisoners as a Vulnerable Research Group

Prisoners are deemed vulnerable because they live in a controlled environment with limited opportunity for making choices, earning money, communicating with those outside the prison, or receiving medical care. Prisoners can participate only in certain types of research, and their consent must be knowing and voluntary.

Pregnant Women, Human Fetuses, and Neonates as Vulnerable Research Groups

The review board must determine if the project is concerned with the mother's health or the fetus's health, and what risks the research poses to the mother, the fetus, or the infant. The researcher cannot be involved in termination of the pregnancy or in determination of the viability of the neonate.

Children as a Vulnerable Research Group

Children are individuals under eighteen years of age. Those who are thirteen to eighteen years old will read and sign, if they agree to it, the informed-consent document. Those seven to twelve years old will read and sign, if they agree to it, the assent document. In addition, a parent or guardian of a child under eighteen years of age will read and sign, if they agree, the informed-consent document.

The review board will approve research that does not involve more than minimal risk to the child. They will also approve research involving more than minimal risk that has the prospect of direct benefit to the child. Sometimes the review board will approve research that has more than minimal risk without direct benefit to the child if it will provide valuable information about the child's disease.

Students at the Institution as a Vulnerable Research Group

The review board would not want students to be coerced into participation in research by their professors who are also researchers. This is especially a concern if research participation is a course requirement. The review board has several ways to decrease the possibility of coercion, including:

Using printed advertisements to recruit students

Avoiding any personal solicitation of students to participate in research

Offering numerous research projects to choose from

Allowing alternative ways (attending seminars, writing a paper, conducting own research, etc.) to satisfy the research requirement

Colleagues at the Institution, Company, or Agency as Vulnerable Subjects

Like students, employees may be subject to coercion by their colleagues or supervisors. The same procedures recommended for use with students are followed. However, the review board does not allow members of the researcher's lab or office to participate.

PAYMENT FOR PARTICIPATION IN RESEARCH

Coercion could also be seen in the use of payment as an incentive to participate in research. The review board considers financial recruitment as compensation for missed time from work, travel, parking fees, and babysitters.

EMERGENCY CARE AND COMPENSATION FOR INJURY AS THE RESULT OF PARTICIPATION IN RESEARCH

The review board ensures that a participant injured in the course of research will receive medical treatment for any injury or illness that is a direct result of participation in research, unless such participation was for the direct benefit of the participant.

FDA APPROVAL REQUIRED FOR RESEARCH ON A NEW DRUG OR A NEW MEDICAL DEVICE

The review board requires that the investigator obtain approval from the Food and Drug Administration (FDA) for research involving a new drug or device. Approval is also necessary for FDA-approved drugs that are being studied for new purposes or doses. No research can proceed without FDA approval.

QUESTIONS TO BE ANSWERED IN THE HUMAN SUBJECT PROTOCOL

The human subjects review board needs to know how the research will be supported. The investigator should include copies of any grant proposal(s) with their application to perform human subjects research. The investigator also needs to indicate any conflicts of interest for themselves or members of their immediate family. Basically, would they or any member of their immediate family receive benefits from the successful completion of the proposed research?

Conflict of Interest in Human Subjects Research

Dr. Jones discovered a gene that produced a protein that appeared to be effective in the treatment of one type of leukemia. Dr. Jones's institution applied for a patent for a therapy based on the protein. A pharmaceutical company licensed the protein and began human subjects research to test the effectiveness of the treatment. Dr. Jones received a grant from the pharmaceutical company and was the lead investigator on this research. When Dr. Jones applied to the institutional review board for approval to try the treatment on patients with leukemia, the board insisted that Dr. Jones disclose the conflicts of interest in this research: Dr. Jones and the institution would profit if the therapy was successful, and Dr. Jones and the institution had received grant funding from the pharmaceutical company manufacturing the treatment. These conflicts of interest had to be included in the informed-consent document so that potential participants could decide for themselves if these conflicts of interest would keep them from agreeing to join the study. In addition, the review board required a consent monitor to be present when the study was explained to potential subjects. The role of the consent monitor was to certify that the participants understood the risks of the study and the possible conflicts of interest.

Lay Language Summary

The investigator must provide a summary of the proposed research in terms that can be understood by all the individuals on the review board, including the community members. There can be no technical terms that could confuse any of the members. This statement provides the context for the study and describes the procedures to be followed.

Purpose of the Study

The review board needs to know the purpose of the study. It needs to weigh any risk to participants against the eventual goals of the research.

Background for the Study

Like a journal manuscript or a grant proposal, the investigator needs to describe in a logical fashion the previous research in the area. The review board needs to understand the context of the study. The researcher needs to show that the proposed study is the next logical step toward solving a particular problem. Like any other literature review, this section needs to include the references cited. The review board wants to be sure that all possible risks are being considered, including those seen in research on nonhuman species.

Number of Participants

The investigator needs to indicate the proposed number of subjects. If there is an existing body of research in the area, the researcher may be able to perform a power calculation to determine the number of participants necessary to determine a difference between groups of subjects at the selected level of significance. In the absence of this literature, the investigator can base the number on the number of individuals in similar research studies. The review board may suggest more or fewer participants based on the members' experience.

Criteria for Inclusion or Exclusion of Participants

In the past, researchers excluded women, children, and ethnic minorities from research. Women were excluded supposedly because of their childbearing potential and the potential impact of the research on their fetuses; however, this does not explain why postmenopausal women were excluded. Children were excluded because researchers wanted to protect them. The rationale for exclusion of underrepresented minority groups is unclear. However, the results of excluding all of these groups have been devastating, resulting in health care disparities. Review boards now attempt to ensure that all relevant groups are represented in the initial application and any renewals.

High-risk studies typically involve individuals who tried all standard therapies without benefit. The rationale is that they need access to novel potential treatments in spite of the risk. These individuals are extremely brave and selfless in agreeing to try an unproven regimen with significant risk. The review boards support them in their quest as long as they understand the potential side-effects.

Studying Mildly Affected Patients

Some genetic disorders have several forms, including an infantile severe form that presents during infancy and results in death, and a juvenile mild form that presents during childhood or adulthood and can be treated without diminishing length of life. Dr. Smith is an expert in one of these disorders. Based on the current understanding of the disease, Dr. Smith wants to try an experimental drug to control the symptoms. Dr. Smith's colleague, Dr. Johnson, is an ethicist who is concerned that the parents of infants with the severe form cannot make a rational decision regarding the participation of their child in the research because of the impending death of their child. Dr. Johnson persuades Dr. Smith to study adults with the juvenile form of the disorder. When Dr. Smith submits his protocol to the review board, it insists that the study include only infants with the severe form so that if there is a benefit to the treatment, they will benefit. The review board dismisses Dr. Johnson's argument about the coercive nature of asking parents to participate who know their child is going to die soon. The review board is acting on behalf of the affected infants and their parents. They are also concerned about the mildly affected patients risking severe side-effects for a potential treatment they don't need.

How Will Potential Participants Be Identified and Recruited?

The investigator needs to specify the relevant characteristics of the target subject for the research. Then the researcher needs to describe how these types of participants will be identified and recruited without coercion. The review board often recommends the use of fliers to recruit subjects. Fliers contain information about the study and how to contact the investigator. If the potential subjects are patients with a particular disorder, the researcher may contact physicians and ask them to explain the study to their patients with this disease and ask them to contact the researcher if they are interested in learning more about the research. In general, the investigator is not allowed to solicit the subjects personally.

What Will Happen to Those Who Participate in the Research?

The investigator needs to explain to the review board exactly what will happen to individuals who agree to participate in the research. This begins with the circumstances of obtaining informed consent and ends with what kind of information the participant will receive at the end of the study. The researcher needs to describe what will happen to the data and any blood or tissue from the study. The review board needs to understand the system for maintaining confidentiality of personal information obtained from participants.

What Are the Potential Risks That Subjects Face?

The review board needs to understand any risks potential participants may face. These risks are determined by previous research on animals and with humans. Those who agree to participate in research need to do so with the understanding that there may be significant risks as well as potential benefits. Each person needs to weigh the risks and the benefits and decide for themselves if it is worthwhile for them to become a research subject.

Nondisclosure of Primate Research Results

Dr. Stevens was proposing research on a new implantable device. Previous research in primates had shown some problems with the device, but Dr.

Stevens believed in it and wanted to make it available to human patients. Dr. Stevens did not include the complications in the prior primate research in the application to the review board. Once the study was approved and Dr. Stevens began implanting the device in human subjects, several of them suffered the same complications as the primates. The review board stopped the study because of these complications. Dr. Stevens explained to the review board that these complications had been seen in primates. The review board moved to bar Dr. Stevens from human subjects research. Similar action was taken by the FDA, which must approve new devices before they can be made available for use.

The review board will work with the investigator to minimize the risks faced by participants in research. It will also seek to maximize the benefits received by participants.

Research without Benefit to Participants

Dr. Clark is a researcher at a government agency. Dr. Clark's research requires large numbers of subjects, repeated blood samples taken from them, and ongoing access to their medical records. Dr. Clark proposes to conduct the research, but not to provide any individual or overall results to participants. Basically Dr. Clark is asking to exploit participants to facilitate the research. The review board requires Dr. Clark to provide regular summaries of overall results to participants who are interested. In addition, Dr. Clark is required to provide individual subjects with results that could significantly impact their medical care, as long as those results meet federal criteria for distribution to individual subjects; for example, they cannot include research laboratory results if that laboratory has not been approved under the regulations established by the Clinical Laboratory Improvement Act.

For research on therapy, the investigator must provide a list of established therapeutic alternatives. Before agreeing to participate in research, potential participants need to know if there is a choice of effective therapy that they may not have considered.

Payment to Participants Needs to Be Explained

The review board will not allow reimbursement to subjects that is so large as to be coercive. The investigator will be required to reimburse participants for their transportation to participate in research, for their childcare expenses, and for their time. Since therapeutic research is designed to determine if there is any benefit, there may not be any benefit, and there may be significant harm in terms of adverse side-effects. If subjects would have to pay for participating in the research, this needs to be explained and justified to the review board.

The review board requires a statement regarding the provision of emergency treatment for the consequence of participation in research. If a drug company is sponsoring a study to test a new drug, or if a device company is researching a new device, the company is responsible for paying for necessary care. Otherwise, the institution, agency, or company carrying out the research may be responsible.

Ensuring That Subjects Have the Capacity to Consent

Individuals who will participate in research need to understand the research and the possible consequences to them. It is up to the researchers presenting the forms to determine the capacity of the participants, parents, or legal guardians. The researchers need to be willing to wait while the potential participant discusses the research with their family and friends. The researchers also need to be willing to answer any and all questions posed by potential participants.

TRANSLATING THE PROTOCOL
INTO THE INFORMED-CONSENT FORM

All aspects of the protocol need to be clearly explained in the informed-consent document. All technical terms need to be translated into everyday language that can be understood by a layperson. This language must be further simplified for the assent forms. Your review board will have specific guidance regarding the format and language for these forms. The goal is to be sure that the participant understands what to expect.

If the participant does not speak English, the review board may require a certified translation into the language understood by the participant.

CHAPTER 16

Research with Animals

Just as human subjects are protected by the review board, there is a similar committee for the protection of animals in research. Unlike humans, animals do not have the capacity to consent and do not have guardians to represent them in this process. The committee serves this role to ensure that animals do not suffer needlessly in the course of research.

ROLE OF THE ANIMAL RESEARCH COMMITTEE

In addition to reviewing research with animals, the committee is also responsible for their care and for training investigators in the techniques of animal research. In addition, the committee requires anyone performing research with animals to complete an annual medical history questionnaire to protect humans and animals from infecting each other. If the investigator removes the animal from the committee-supervised environments and keeps the animal for more than twelve hours, the committee is required to inspect these areas, and the investigator is required to maintain accurate records on activities with animals in these areas. If the investigator is performing survival surgery in these areas, there are additional requirements. Each institution, agency, or company conducting animal research will have its own committee.

ROLE OF THE VETERINARIANS

The veterinarians are responsible for the health and safety of the animals under their care. When a new shipment of animals arrives, the veterinarians are responsible for examining these animals, treating those who are ill, and quarantining the new animals. They also set up breeding pairs for investigators who have committee approval to have a breeding colony. They monitor the health of the animals in their care and notify investigators of any health problems or signs of distress. The veterinarians permit access to the animals only by investigators with approved, up-to-date protocols that cover the procedures described in the protocols.

The veterinarians are responsible for updating policies regarding blood collection, tail biopsies, toe clipping, anesthesia, euthanasia, surgery, and analgesia. They encourage investigators to consider alternatives to the use of animals in research. They provide instruction in and monitoring of surgical techniques.

INFORMATION IN AN
ANIMAL RESEARCH PROTOCOL

Before initiating animal research, an investigator must have a protocol approved by the committee. The protocol begins with a summary of the proposed research, including the rationale for the use of animals and the experimental design. It includes the personnel involved and their contact information. Each individual must have completed the requisite training to be involved with the protocol. The investigator needs to indicate the source of support for the research.

In an attempt to minimize discomfort to the animals, the investigator indicates the pain category and analgesia to be used. Any anesthesia and surgery need to be described. The investigators also need to specify where the animals will be housed. The investigator assumes responsibilities for the actions of the members of his or her lab group for their animal research.

Pressure on Lab Members

Dr. Park is a major researcher who is known for being hard-driving. Dr. Park's graduate student Kim is feeling a lot of pressure to complete some research with animals in response to reviewers' comments on a recently submitted manuscript. The journal editor asked Dr. Park to complete these experiments and then resubmit the manuscript. Kim knows not to start these procedures until they have been approved by the animal research committee. Dr. Park knows this too, but is constantly asking Kim when the experiments will be done. Kim bows to Dr. Park's pressure and starts the research without committee approval. When the veterinarian observes Kim performing these procedures, all of Dr. Park's animal research is shut down until a full review can be completed.

As with the human subjects review board, the investigator should accept the recommendations of the animal research committee. Many of its recommendations are required by the Association for Assessment and Accreditation of Laboratory Animal Care for accreditation for animal research.

CHAPTER 17

Data and Sample Management

Your responsibility to your human and animal subjects extends to how you deal with the data and tissue samples obtained from them. For human subjects, you need to maintain confidentiality of data and to use the samples for the intended purpose. For animal subjects, you need to use the samples appropriately with the goal of minimizing the number of animals tested. All research data need to be kept with the principal investigator in perpetuity, unless the investigator leaves the institution. Research data and samples belong to the institution. An investigator who is leaving the institution where research was performed must formally request transfer of the data and samples to the new institution. Such permission is generally, but not always, granted.

OWNERSHIP OF DATA

Granting organizations, such as the National Institutes of Health, have a contract with the principal investigator and the recipient organization that requires raw data in the lab notebooks and computer files to be stored for at least seven years so that it will be available for review. Similarly, an oversight organization, such as the Food and Drug Administration, may require a review of original data. In addition, journal publishers may require that data be maintained indefinitely in case a challenge to a journal article is made. Our journal recently had a challenge to a paper published twelve years ago.

Original lab notebooks and computer files (with backup) need to be maintained in the research environment. They cannot be changed arbitrarily. Changes in a paper notebook must be dated, signed, and justified. The original data must be maintained. Electronic copies of data should be dated and stored on a CD-ROM.

Redundancy Is Important

Taylor was writing her dissertation. She wanted to focus on her writing at home over the weekend. She put her laptop and lab notebooks in the trunk of her car and headed home. On the way home, she was carjacked. Although she was not harmed, her five years of hard work were in the hands of the carjackers. She immediately contacted her mentor, Dr. Stevens. Dr. Stevens could see that Taylor was very upset. Dr. Stevens was disappointed that Taylor had ignored the rule of the research group to leave lab notebooks in the lab at all times. Fortunately, Taylor had already submitted three papers based on her dissertation research, and Dr. Stevens was working on a grant proposal that contained original data and methods from the remaining section of Taylor's dissertation. Between the electronic versions of the manuscripts and the proposal, Taylor was able to reconstruct her dissertation.

CONFIDENTIALITY OF DATA

As part of the informed consent with human subjects, the investigator agrees to maintain confidentiality of personal information obtained from the participants and any individual results from the research. When an individual agrees to be a part of a research project, they are given a code number and all of their personal information, samples, and data are stored with this code. The investigator maintains the key to the codes and identities in a locked cabinet so that only the investigator and possibly one other person can access the key. Computerized data are coded with the key password protected. Other researchers know participants solely by their code.

Another way to maintain confidentiality is to strip all identifiers upon receipt of the individual's information, sample, and data, and code the material. With this anonymous material, there is no chance for breach of confidentiality. However, there is no possibility of linking new information or samples to the original individual. This process prevents sharing of individual results with participants.

The approach to confidentiality needs to be clearly spelled out in the informed-consent document so the potential participant understands exactly what to expect in terms of feedback. Participants need to know whether they will receive group results, individual results, or no information.

SHARING OF DATA SETS AND SAMPLES

Research subjects also need to understand from the informed-consent document what will happen to their personal information and/or sample. Will these just be used for the single study covered by the informed-consent document? Will they be used for research by this investigator and others on the topic of the current study? Will the personal information and/or sample be shared with other researchers studying other topics?

Before other investigators are given access to subject information and/or samples, the other researchers need to know the stipulations of the informed consent for the original study. Under no circumstances should information and/or samples be shared with other investigators or for other purposes unless explicitly stated in the consent document. If circumstances change and the original researcher would like to share information and/or samples though this was not part of the informed consent, then the original researcher needs to amend the human subjects protocol and informed-consent document and must recontact the participants. The original subjects should not be coerced into agreeing to the terms in the revised informed consent.

CONTROVERSY REGARDING SAMPLE OWNERSHIP

Institutions seem to regard research samples as their property, while the National Institutes of Health considers investigators as stewards of the samples and expects researchers to manage the samples so as to provide the most benefit to the human subjects.

Three legal cases support the point of view of the institutions. *Moore v. Regents of the University of California* concluded that Mr. Moore had ceded ownership of his samples to the University of California in spite of the fact that he had not signed an informed-consent document for research on his samples. *Greenberg v. Miami Children's Hospital* determined that the patients and their families did not have a right to the patent discovered using their samples. *Catalona v. Washington University* found that Dr. Catalona could not move his research samples with him to Northwestern University to continue his research, in spite of the fact that 6,000 of the participants in his research asked that their sample be returned to them from Washington University so their sample could be given to Dr. Catalona at Northwestern University. They had signed informed-consent documents that stated that they could withdraw from the research at any time.

MANAGEMENT OF TISSUE BANKS

Collections of specimens for research purposes falls under the supervision of the human subjects review board if they are from humans, and under the animal research committee if they are from animals. The human subjects review board has a special interest in how the human tissues are used. Anyone with a human tissue bank must have an approved protocol with the review board before samples can be shared with other investigators. Any investigator attempting to access a tissue bank's samples must have an approved protocol with the review board. Managers of human tissue banks need to have a process for determining which approved protocols can have access to the valuable, finite resources.

REQUIREMENTS OF FUNDING AGENCIES AND JOURNALS

Funding agencies, such as the National Institutes of Health, require investigators to share reagents and data developed with grant support. In addition, publications of grant-supported research need to be made available by the publisher to the National Library of Medicine.

Depending on the research, journals may have specific requirements for data sharing before an article can be published. The authors need to provide database access information as part of the Methods section of their manuscript before publication can proceed.

This sharing of resources is critical for the advance of science, but must be carried out according to the requirements of the review boards.

CHAPTER 18

Ethical Behavior

Every vignette in this book represents an ethical dilemma that one may face in the course of developing an academic career. Typically the answer has been to seek advice from your mentor(s). You might ask if academia is especially prone to these issues. The answer is that whenever people are involved, there are similar problems. With your mentors you can confront these issues and establish your own ethical standards. As a mentor yourself, you will need to live by these principles as a model for your mentees.

DEVELOP FUNDAMENTAL PRINCIPLES FOR DEALING WITH ETHICAL DILEMMAS

During your education and training you will receive formal and informal instruction on ethical behavior. All NIH-funded training programs require that all trainees participate in a formal course on ethical behavior. In addition to coursework, you should also observe those around you. Mentors provide training in ethical behavior through both exemplary conduct and discussions of dilemmas they or others have faced.

For those who find themselves in a difficult position, many academic settings provide ombudspersons to deal with ethical concerns in a confidential manner. An ombudsperson is skilled in analyzing the situation and helping the individual identify a solution.

You must recognize that for most ethical dilemmas there is no one single correct solution. Therefore, it is important for you to develop your own fundamental concepts and principles with which you will be able to consider dilemmas as they arise. The following examples are written to stimulate you to consider some of the ethical issues that you may face during your academic career. Your mentors and your institution's ombudsperson can provide advice when you face such dilemmas.

ETHICAL SITUATIONS

Ethical Situation 1: You Have a Right to Information Concerning Your Own Career

Kim was enjoying an academic career and felt successful during the first year. At the end of each year, the department chair and the associate dean for academic affairs invited each faculty member to discuss the past year and receive feedback. Kim thought the meeting with the department chair went very well, so well that Kim did not ask to see the report the department chair forwarded to the associate dean. Kim's meeting with the associate dean did not go well at all. It was as if Kim were two different people. Fortunately, Kim went to a senior faculty member to discuss the discrepancy. The faculty member asked if Kim had read the department chair's report on Kim's performance. When Kim replied no, the faculty member suggested that perhaps Kim should make another appointment to meet with the associate dean. During the second meeting with the associate dean, Kim asked to read the department chair's report. The report was extremely negative. The associate dean offered to meet with the department chair to discuss the discrepancy. The associate dean suggested that the department chair had preferred another candidate for Kim's position and that perhaps the chair wanted to encourage Kim to leave so that the other candidate could be hired.

It is very important for you to review the information in your files that are open and available to you. You may find discrepancies with your own impression of your progress. These insights will help you to improve. There also may be outright errors that you need to correct. If you feel that there is a discrepancy between the verbal and written comments about you, you need to try to clarify the discrepancy. There have been cases of written documentation of performance being used both to encourage people to leave or to force them to stay by limiting their opportunities to leave.

Ethical Situation 2: Do Not Agree to Conditions Unless You Can Comply

When Lee joined Dr. Smith's research group as a postdoctoral fellow, Lee agreed that his project would remain with Dr. Smith when he left the group. This was standard practice in Dr. Smith's group. All trainees made the same agreement before they began to work with Dr. Smith. Two years later, Sam, a graduate student in Dr. Smith's group, finds Lee furtively copying his research files on the eve of his departure for his first job. When Sam asks Lee the purpose of his copying, Lee replies that he plans to continue his project at his new position. Sam reminds Lee that they had all agreed not to take their research with them when they left. Lee replied that he had changed his mind and any agreement with Dr. Smith was bogus. The first thing the next morning, Sam speaks to Dr. Smith about Lee's behavior.

You need to understand that funding agencies consider the research to belong to the institution in which it was performed, and that the institution delegates this responsibility to the principal investigator. It is distressing, however, when a mentor, by keeping control of the entire project, does not help a trainee move to the next level. Most areas are broad enough to permit a trainee some opportunity to continue their momentum. On the other hand, any trainee who has entered into an agreement must honor it. While there is probably no legal foundation for the agreement, the trainee will need letters of recommendation from their mentor. The trainee who has made such an agreement should be considering and even developing alternative projects in the course of their training that they will be able to take with them.

Ethical Situation 3: Sexual Harassment and Other Forms of Power Abuse Should Not Be Tolerated

Stacey, a young faculty member, finds a student, Lee, to be very upset. When Stacey asks Lee what the problem is, Lee replies that Dr. Jones (a very senior faculty member in Stacey's department) has demanded sexual favors in return for awarding the grade Lee earned in Dr. Jones's class. When Lee told Dr. Jones to "bug off" and threatened to report Dr. Jones to the department chair, Dr. Jones replied that Dr. Jones would see that Lee would never be able to pursue graduate studies. Stacey suggests that they both meet with the department chair to discuss Dr. Jones. Lee refuses due to concern about a future career. Stacey goes to the chair of the university Committee on Sexual Harassment. The committee chair assures Stacey that their inquiry would protect Lee's identity. When Stacey next speaks to Lee, Lee is willing to meet with the committee. Dr. Jones had solicited sexual favors from a number of students in exchange for grades. Several students agreed to go to the committee. As a result of this inquiry, Dr. Jones retired immediately and was unable to interfere with the future careers of the students who testified before the committee.

Sexual harassment, which may occur with any gender combination, is unconscionable and illegal. If the victim is under any administrative control by the abuser, and if the institution has not set up a spe-

cific mechanism (e.g., an ombudsperson) to deal with this problem or other examples of power abuse, the victim needs to identify someone with sufficient sensitivity and security to take on this issue. In the absence of such an individual at the institution/institute/agency, the victim should consult a lawyer and consider legal action.

Ethical Situation 4: Results of Investigations Should Be Published

Chris, a graduate student, worked on a research project with a postdoctoral fellow. The fellow took the lead on the project, and was first author on an abstract submitted to a professional meeting. Chris was second author on the abstract. The fellow left for another position. Chris asked the former fellow repeatedly about writing up the manuscript. The former fellow claimed no interest in doing so, because the work was irrelevant to the fellow's new position. Chris went to the research group leader and offered to write the manuscript and be the first author. The group leader agreed that Chris should assume this role.

Research does not count if it is not published. The resources involved in obtaining the results are enormous. When some individuals leave, they may lose interest in publication. However, the others involved still need the publication. Someone else can step in to facilitate writing up the project and should then become the first author. The original first author then moves to a middle authorship position.

Ethical Situation 5: Review of Human Subjects Research

Kelly was applying to graduate school while doing research with Dr. Smith, a senior faculty member. Kelly was impressed by all of the patient samples that Dr. Smith had collected over the years. When he was contacted by a colleague with a request for a sample from a patient, Dr. Smith was always happy to send it along. Dr. Smith continued to add

to his collection of patient material. Kelly asked Dr. Smith if the patients had agreed to participate in his research, and he replied, "Of course." Kelly remained curious and asked if the research project was explained to the patients. Dr. Smith replied that there was no need to bother his patients with details since the research was for their benefit. Still not satisfied, Kelly went on the university website and found the office for human subjects research. Kelly learned that Dr. Smith should have approval for human subjects research from this office, an informed-consent document for participants, and a patient sample storage protocol. Kelly confirmed this with one of her former professors. They both went to talk with Dr. Smith to urge him to conform to the rules for human subjects research.

During an investigator's career, the rules may change. Continuing to ignore the rules for human subjects research puts the investigator and their research and samples in jeopardy. In this vignette, thanks to the insight of an undergraduate, the investigator learned he needed to complete the appropriate forms and receive approval for his human subjects research. Through this process, future participants would complete the informed-consent document at the time of enrollment. The investigator would need to contact previous patients. If they agree to participate and sign the consent form, their sample can be studied. If they do not agree or cannot be located, the investigator cannot use the sample in the research. This process is required to continue the research and to publish the results, since journals all require approval by the human subjects research office of the author's institution.

Ethical Situation 6: Oversight of Research with Mice

Kim's group was in a very tight race with another research group to be the first to show successful gene therapy in mice that would prove to be

a new model for treatment of a human disease. His mentor, Dr. Jones, had approval for gene therapy research from the institution's animal research committee. So far, the approved gene therapy protocol was not successful. Kim read some articles that suggested a different route of administration. Dr. Jones thought that they should try the new route. Kim replied that they would need to get approval from the animal research committee and training from the institution's veterinarian before they tried the new method. Dr. Jones became really angry and told Kim to just do it. Dr. Jones said, "We don't have time to waste on updating the protocol."

In such a case, the trainee needs to talk with a senior mentor who performs animal research and enlist their help. The principal investigator would, with such rash demands, be placing the safety and welfare of the research animals, and the entire research project, in danger.

Ethical Situation 7: Avoid Overinterpretation of Data

Sandy met with her mentor to present some new data. Her mentor was excited by her results and suggested she reanalyze the data to emphasize some small differences between the two groups. Sandy was very concerned about this request and met with the department chair to discuss these issues. The department chair praised Sandy and suggested that they meet with her mentor to prevent overinterpretation of her data.

Situations like the one above may represent overinterpretation of results to fit the mentor's favorite model. On the other hand, the mentor may have insight into the results and methodologies that permits an interpretation not understood by the mentee. In such situations, it is critical for the mentee to get a senior person involved and ask them to review the results and provide their interpretation.

Ethical Situation 8: Selecting Data to Support
Your Hypothesis Is Not Acceptable

Kim, a postdoctoral fellow working with Dr. Jones, collected a large number of subjects and controls for a study. Kim performed the data analysis and found no statistically significant difference between the subjects and controls. Kim brought the raw data and the analysis to a meeting with Dr. Jones, Kim's research mentor. Dr. Jones reviewed the raw data and the data analysis and suggested that Kim remove a number of subjects with "outlier" values. When Kim questioned Dr. Jones, Dr. Jones said that if Kim didn't do this, Kim's work would not be accepted for presentation at the next professional meeting or for publication.

In such a situation, the trainee clearly needs an additional mentor to discuss the one mentor's suggestion that the trainee commit what may be scientific fraud. This person should be another scientific mentor or someone more senior in age or rank to the investigator. The three of them should meet and discuss the situation and develop plans for the trainee's future research. These plans may include the selection of a new scientific mentor. Graduate students have dissertation committees to oversee and review their research. Oversight committees for postdoctoral research can prevent the misuse of data and methods.

Ethical Situation 9: Honesty Is an Absolute
Requirement in Research

Chris and Stacey are postdoctoral fellows in the same group. Chris's research has been progressing very well. Chris already has two papers in top-level journals. Stacey's project is moving much slower. To help Stacey, Chris often reviews Stacey's data. Chris is surprised when Stacey presents data in a group meeting that are the exact opposite of the data Stacey has been getting week after week and showing to Chris. The data

Stacey presents at the group meeting send their mentor into ecstasy. These data are just the thing to support a major aspect of the mentor's favorite model. When Chris asks Stacey what changed to get the new results, Stacey brushes Chris off. Chris is concerned that Stacey made up the results to please their mentor. Chris meets with their mentor to discuss these issues. When the mentor confronts Stacey, Stacey confesses to making up the data and is dismissed from the research program.

Your credibility in your field is based on your reputation. Fabrication of data is absolutely unacceptable. We must be completely honest because research is based on trust. You earn the respect of your peers through your skills. People know you by what you do, not by what you say. Your reputation is hard to establish, but easy to lose. You should seek to be trustworthy, have integrity, and be respected.

Ethical Situation 10: Faking a Theft Has Real Consequences

Kim is a graduate student who has not made any progress over several years because Kim has not done any research during this time. Kim realizes that folks would notice the lack of progress. Kim reports that data notebooks and computers have been stolen. Since Kim's work was in an area involved in national security, the theft comes to the attention of federal authorities. The university determines that Kim has not been performing research and that Kim's recent paper is based on falsified data. Kim has to retract the paper. Kim also has to repay the research grant funds that have been used for salary support, research supplies, and travel to professional meetings. Kim is sentenced to prison for faking the theft and lying to federal authorities.

None of us is smart enough to lie and get away with it. In this vignette, the trainee had been lying about data—made-up data for a manuscript—and eventually realized that someone would want to see the raw data. Lying about the theft of the data did not help in this case because of the national security implications of the work. However, nearly all scientific fraud is eventually uncovered when others are unable to replicate the fraudulent data. The trainee would have been better off doing the work in the first place instead of lying about it.

Ethical Situation 11: Reviewer Misdirecting a Paper under Review

Dr. Johnson is asked to review a manuscript for a prestigious journal. She agrees to do so since the manuscript involves a new treatment for epilepsy that she previously studied for the company that produces the drug. The manuscript shows a lack of effectiveness and significant side-effects of the medication. She sends a copy to the person at the company who was in charge of her grant to study this treatment for the company. Should a reviewer share a copy of a manuscript?

No reviewer should share a copy of a manuscript. What was the investigator expecting from the company for sharing this manuscript? Clearly the investigator's behavior showed she had a conflict of interest. She wanted to improve her relationship with the company and used the copy of the manuscript to do so.

Ethical Situation 12: Self-Interest Guides Review of a Colleague

Dr. Jones is being considered for a promotion. Dr. Smith is a member of the committee that evaluates promotions for the department. Dr. Smith needs space and would like to take over Dr. Jones's research space. Dr. Smith offers to prepare the committee report for Dr. Jones's promotion. Dr. Smith describes Dr. Jones's research group as small, Dr. Jones's research publications as appearing in second-tier journals, and Dr. Jones's research as "good." The committee accepts Dr. Smith's report, and Dr. Smith pre-

sents the report to the departmental faculty meeting. At the meeting, Dr. Johnson, a research collaborator of Dr. Jones, challenges Dr. Smith's report. Dr. Johnson notes that there are twenty people in Dr. Jones's group, hardly small for this research area. Dr. Johnson notes that Dr. Jones's work has revolutionized the field and suggests that Dr. Smith check the number of citations of several of Dr. Jones's key papers, as well as the impact factor of the journals in which Dr. Jones publishes. Dr. Johnson considers Dr. Jones's work to be not good, very good, or excellent, but outstanding. Dr. Johnson and the department chair suggest that Dr. Smith and the committee prepare a more accurate description of Dr. Jones's research. The department chair also refers Dr. Smith to the committee on faculty conduct for censure for attempting to impugn Dr. Jones's research in order to appropriate Dr. Jones's research space.

There are consequences for the behavior of the faculty member who provided the inaccurate review, which was guided completely by self-interest. While the participation of faculty in the evaluation of one of their peers is valued for self-governance, it is not an opportunity to ruin a colleague's reputation. Fortunately, in this case, another faculty member and the department chair rose to the defense of their colleague. If they had not, the errant faculty member may have succeeded in discrediting the faculty member under review and obtaining their research space. One has to wonder if the reviewing faculty member's willingness to distort another faculty member's research might be seen in the reviewer's approach to their own research.

Ethical Situation 13: Commitment to a
Percentage Effort Level on a Grant

Kim is supported by an NIH postdoctoral training grant that requires 80 percent effort on research. Kim's division chief agreed to this before Kim accepted the funding. Now the division chief is constantly pressuring Kim to spend 60 percent time on clinical duties, leaving only 40 percent time for research. Kim's mentor meets with Kim's division chief and reminds the division chief that acceptance of Kim's salary from the NIH commits the division to honor the 80 percent effort on research. Kim's mentor threatens to take the issue to the department chair and the university office for research if Kim's clinical effort is not reduced to 20 percent. Faced with Kim's mentor's ultimatum, the division chief reduces Kim's clinical effort to 20 percent.

A contract is only as good as the two people signing it. A trainee's division chief who does not honor the contract bears watching, since the best predictor of future behavior is past behavior. The trainee needs to establish a good relationship with the department chair and several other senior mentors if the trainee stays at this institution on the faculty. It is highly likely that a division chief like this one will go back on other agreements as well. The trainee will need advice on how to deal with these indiscretions. The department chair and other senior members of the department need to be aware of the division chief's behavior.

Ethical Situation 14: Comings and Goings

Lee is leaving one research group to join another. Before she leaves, Lee spends a lot of time complaining about her current situation. She also builds up the new group as the best possible opportunity. When Lee joins the new group, she continues to complain about her former group. Her new colleagues wonder how she could have chosen her previous position and how she could have been with them for so long. They also wonder what Lee really thinks of them. Will she have unkind things to say when she leaves their group?

No relationship is perfect. This individual probably had very good reasons for selecting the first research group and for staying with them for so long. In trashing this group she is diminishing herself for her long-standing group membership. Her new colleagues have to wonder if she will have the same opinion of them when she leaves the new group.

Ethical Situation 15: Power Abuse

Dr. Jones is teaching a course for graduate students and spends much time criticizing the graduate students, commenting on the inadequacies of their generation as a whole. All the students in the class are uncomfortable with Dr. Jones's behavior. They are, however, afraid of confronting Dr. Jones directly, because Dr. Jones has a reputation of being very vindictive. Dr. Jones becomes increasingly strident as the course goes on, making the students even more upset.

While everyone is entitled to their opinion, no faculty member has the right to express negative feelings toward a group of students possessing a particular personal characteristic. The students have a number of individuals they should approach regarding their concerns, including the director of the graduate program, department chair, dean of the graduate school, university ombudsperson, provost, and/or president/chancellor. It is a privilege to teach. A professor who denigrates students should not be afforded the opportunity to do so again, particularly when there is a history of similar behavior.

Ethical Situation 16: Negotiating a Contract

When Chris negotiated an initial faculty appointment, everything was put in writing by the department chair. Now Chris finds that the chair is not living up to half of the promises made in the offer letter. Chris understands that some things are beyond the chair's control. For example, Chris's new office was not ready on time. Chris realizes that the chair is not addressing other issues, such as the promised research space that was given to a more senior faculty member as part of a retention package by the time Chris arrived. Does Chris have any recourse?

The new faculty member should discuss the discrepancies between the offer letter and reality with the department chair. If this does not resolve the issues, there are further options. Did someone else sign the offer letter besides the department chair, such as perhaps the dean? If so, the new faculty member should meet with the cosignatory to seek their support. This faculty member should also discuss this with their research mentor and with their mentor for promotion and tenure. All of these individuals should bring pressure to bear on the department chair with the goal of getting the chair's renewed commitment to the offer letter, and a revised but speedy time table.

Ethical Situation 17: Dealing with Plagiarism

Stacey requires an original paper from each student. Two of the students submit papers that are word-for-word identical. Does this mean that one student misappropriated another student's paper? Or did both students purchase the same paper from a ghost writer? Or did both students download the paper from the Internet or copy text from a journal or book?

Faculty members are requiring students to submit their papers for online plagiarism detection. This is essential to ensure meaningful grades for each student in the course. The consequences of plagiarism are severe. Typically the student receives a grade of F or zero for representing someone else's writing as his or her own. Some schools have a student ethics committee or honor code committee, to which students who plagiarize should also be reported.

CONCLUSION

As you can see from these ethical situations and the vignettes throughout the book, each has multiple possible solutions. Your mentors have had similar experiences or have witnessed like episodes. They will have strategies to help you cope with these issues. You need to make the best of the situations in which you find yourself. You need to grow professionally and personally and reach out to help the next generation.

The theme throughout this book is the importance of mentors. You should never be embarrassed to go to a mentor if, for example, a manuscript is rejected or a grant is not funded. You should not feel you are wasting a mentor's time when you seek their advice. Our biggest mistakes have been when we neglected to approach a mentor and ask for help.

Mentoring immortalizes you through your mentees. For the most part, the papers you publish today will not be very significant in ten years, but the person you mentor today will carry on your intellectual legacy. What an enduring tribute!

INDEX